Osteoimmunopathology

Francesco Chiappelli

Osteoimmunopathology

Evidence-Based Perspectives from Molecular Biology to Systems Biology

Springer

Professor Francesco Chiappelli, Ph.D.
University of California at Los Angeles
School of Dentistry
Division of Oral Biology & Medicine
Los Angeles, California
USA
and
President, American Association of Oral Biologists
Evidence-Based Dentistry Champion
American Dental Association
Los Angeles, California
USA
fchiappelli@dentistry.ucla.edu

ISBN 978-1-4419-9494-3 e-ISBN 978-1-4419-9495-0
DOI 10.1007/978-1-4419-9495-0
Springer New York Dordrecht Heidelberg London

Library of Congress Control Number: 2011926796

Printed on acid-free paper

Springer is part of Springer Science+Business Media (www.springer.com)

Preface

The Healthcare Bill: Toward Evidence-Based Healthcare

The U.S. Patient Protection and Affordable Care Act (enacted on 23 March 2010) created a $1 billion program to award, on a competitive basis, federal subsidies to qualifying therapeutic discovery projects being conducted at small biomedical companies. Within the first 6 months of its enactment, the new law not only brought the U.S. up to par with the rest of the developed nations in a general sense with respect to guaranteeing the right to affordable healthcare for its citizens, but also more specifically ensured, among many other fundamental implementations, that (cf., http://www.healthcare.gov/news/factsheets/overview.html):

- Insurance companies be regulated so as to prevent unjustified premium increases and to put in place common sense policies.
- Small US businesses be afforded some form of business tax credits to help cover their employees.
- Most uninsured US citizens and legal Residents, who had been left uninsurable due to preexisting conditions such as certain of the immunological and bone pathologies that are discussed herein, now have access to a quality health insurance that will provide them with the necessary coverage for their ongoing and future treatments.
- An Early Retiree Reinsurance Program ensures continued health coverage for US citizens and legal Residents forced to retire early – due to the constraints of the recent international economic downturn – and thus not yet eligible for Medicare.
- Medicare beneficiaries received $250 as assistance to help them afford the cost of prescription drugs in the Part D "donut hole" coverage gap, and a mandated 50% discount on brand name drugs for future purchase – indeed, a substantial financial help for our elderly, most of whom suffer from some form of the osteoimmune pathologies described here, and thus actively under treatment with the pharmaceuticals discussed in this monograph.

Osteoimmunology pertains to the study of the relationship between the bones, particularly the bone marrow, and the immune system. This monograph pursues the best available evidence, by means of research synthesis, for the characterization of the physiological relevance and pathological implications of the interconnectedness between the skeletal and the immune system. Research will be discussed that highlights the associated role of the circulatory, nervous, and endocrine systems, as well as proteomic and genomic pathways and signatures. Emphasis will be given to the implications to stomatology, that domain of medicine which relates to the oral cavity, its diseases, and their systemic *sequelae*.

This monograph arises from observations that have suggested that the skeletal system and the immune system are very tightly bound together. Chronic inflammatory reactions subsequent to an excessive immune reaction can damage the bones, as in rheumatoid arthritis (RA), Ankylosing spondylitis, an autoimmune disease characterized by arthritis, inflammation, and eventual immobility of joints (AS, Bechterew's disease), osteoporosis, with a special emphasis on patients seropositive for the human immunodeficiency virus (HIV)-1 and with signs and symptoms of the acquired immune deficiency syndrome (AIDS), and bone cancer. Bones – in particular the bone marrow – are one of the primary locations in which cells of the immune system mature. The precise ways in which the bones and immune system influence each other have now begun to be understood.

In brief, this monograph seeks to answer the following questions:

- What is osteoimmunology all about?
- How do the immune system and its components affect bone development?
- What influence does bone, particularly the bone marrow, have on immunity, and what influence have immune factors, such as cytokines, have on bone?
- Do stress and hormones impact upon the pathophysiology of bone-immune interactions?
- Can we, through the scientific process of research synthesis, obtain the best available evidence for treatment of diseases involving the bone-immune entity (i.e., osteoimmunopathologies)?
- What might be the stomatological correlations and implications of osteoimmunology and osteoimmunopathology?
- How might we improve certain stomatological conditions and systemic *sequelae* by means of evidence-based clinical decision-making directed at the treatment of osteoimmune pathologies?

Toward Translational Evidence-Based Osteoimmunology: Relevance and Implications for Stomatology

Osteoimmunology (Gr: οστεον,=osteon, bone; La: immunitas, immunity; Gr: λογοσ, logos, knowledge) is the study of the interface between the skeletal system and the immune system, comprising the "osteoimmune system." It is also the study of shared components and mechanisms between the two systems in vertebrates, including ligands, receptors, signaling molecules, and transcription factors. Some medical conditions in which this field is particularly relevant are bone metastases, rheumatoid arthritis, osteoporosis, periodontitis, and other inflammatory diseases of the bone. Studies in osteoimmunology reveal relationships between molecular communication among immune cells and structural physiopathologies of the body.

In brief, the field of osteoimmunology pertains to the complex interactive communication between the immune and skeletal systems, as it pertains directly to the physiology of bone and of immunity, as well as to a wide spectrum of pathological conditions that range from autoimmune, to inflammatory, and to neoplastic diseases. Immune and bone cells share a variety of mutual signaling molecules, growth factors, and signaling pathways, as well as a common site of origin, namely bone marrow, which is rich in blood supply, and autonomic innervation, thus ensuring well-distributed neuroendocrine modulation of bone, immune, and osteoimmune events.

Stomatology (Gr: στομα, = stoma, opening or in the normal anatomy[1] context the mouth) refers to that specialty of medicine that addresses the mouth and its associated diseases, and their oral and systemic *sequelae*. The stoma is an anatomically important and physiologically critical opening that serves both the respiratory and the gastrointestinal systems. The oral cavity opens at the mouth and nares (i.e., nostrils) into the posterior stomatological space, the oropharyngeal isthmus (or *fauces*[2] [Lt, pl, throat openings]). The fauces open into the pharyngeal space, which consists of the nasopharynx, the oropharynx, and the laringopharynx (about the hyoid bone anteriorly, and the fourth to sixth cervical vertebrae). It is in the laringopharynx that divergence occurs between the respiratory (larynx) anteriorly, and the digestive (esophagus) pathways. The laringopharynx is continuous with the esophagus posteriorly.

The osteology of the stoma is complex, and consist of several interconnect facial bones, the bones of the anterior and lower human skull. In brief, these bony structures include:

- *Inferior nasal concha*, which extends horizontally along the lateral wall of the nasal cavity, posteriorly to articulate with the conchal crest of the palatine bone, anteriorly to articulate with the conchal crest of the maxilla.
- *Maxilla*, with its characteristic alveolar processes that hold the upper teeth, and its articulation laterally to the zygomatic bones. In addition, important features of the maxilla are the zygomatic process, the frontal process, the palatine process, the infraorbital foramen, and the maxillary sinus.

---

[1] In pathological surgery, the word stoma (pl, stomata) is used to indicate a surgical procedure that engenders a novel opening (e.g., colostomy).

[2] The fauces consist of two distinct mucous membrane structures that signify the posterior aspect of the oral cavity: anteriorly, the palatoglossal arch, and posteriorly, the palatopharyngeal arch. The fauces guard the organism from invasion of pathogens by means of the Waldeyer ring, named after the German anatomist Heinrich Wilhelm Gottfried von Waldeyer-Hartz (1836–1921), the ring of lymphoid tissue around and about the naso- and oropharynx (i.e., from superior to inferior, pharyngeal tonsils, Eustachian tubal tonsils, palatine tonsils lingual tonsils).

- *Palatine bone*, which forms what is commonly known as the hard palate, and articulates with five bones of the deep face, important for the internal structure of the oral cavity: the sphenoid, ethmoid,[3] maxilla, inferior nasal concha, vomer bones.
- *Vomer bone* is a thin, somewhat quadrilateral bone situated in the median plane, and forms the hinder and lower part of the nasal septum. Along its two surfaces runs the nasopalatine groove obliquely downward and forward to aid the naso-palatine nerve and vessels. Its inferior border articulates with the crest formed by the maxilla and palatine bones.
- *Zygomatic bones* play a critical structural role as they articulate with the maxilla, the temporal bone, the sphenoid bone, and the frontal bone. The malar aspect presents the zygomaticofacial foramen for the passage of the zygomaticofacial nerve and vessels; the temporal aspect supports articulation with the maxilla, and forms the anterior boundary of the temporal fossa, the lower a part of the infratemporal fossa, an irregularly shaped cavity, situated below and medial to the zygomatic arch, and bounded laterally by the ramus of mandible.
- *Mandible* articulates with the two temporal bones at the temporomandibular joints, complex synovial joints (i.e., gliding or arthrodial) whose motion, while considerable occurs in one plane only (i.e., hindge or gingymal), and which suffer from several dysfunctions, including inflammatory osteoarthritis, with significant local and systemic *sequelae*.
- *Hyoid bone* rests at the level of the base of the mandible in the front and the third cervical vertebra behind, and provides attachment to the muscles of the floor of the mouth and the tongue above, the larynx below, and the epiglottis and pharynx behind.

The ossification of these structures in ontogenesis, their articulations in adult life, and related osteoimmune pathologies and evidence-based treatments, are discussed.

Los Angeles, CA Francesco Chiappelli

[3]Both the ethmoid and the sphenoid bones are superior and deep to the oral cavity proper, and serve more properly the nasal, rather than the oral stomata.

Acknowledgments

This monograph arose from many collegial conversations with Professor Jeanne Nervina, DDS, PhD, formerly at UCLA and now at the University of Michigan. Out of our collegiality emerged a graduate course, which we jointly offered as part of the Oral Biology graduate program at UCLA. The course was entitled "Osteoimmunology," and presented the students with balanced materials from Professor Nervina's expertise in bone biology, and mine in cellular immunology. This monograph could never have been possible without Dr. Nervina's superb intellectual contribution, the excellence of her teaching, and the quality of her research in this field. In an earlier format, this monograph was to be coauthored by Professor Nervina and myself, but, due to unforeseen time constraints, she could not continue her contribution to this work and elected to retract her name. It is thus with regret that I had to restructure the monograph below, and present it in its present single-authored format. I can never cease to emphasize, however, that any laud of this work must be attributed to Dr. Nervina, equally as to this author.

Critical intellectual assistance was provided to me in the compilation of the materials describing the fundamental bone biology by my doctoral student, Andre Bardhokian. His constant and preservering efforts in my laboratory have generated and continue to generate excellent novel and cutting-edge developmental bone biology and osteoimmunology knowledge. Andre has developed two courses (one of them web-based distance-learning) on the topic of osteoimmunology, which are presently being offered through the UCLA University Extension. Many of the discussions and brainstorming sessions with Andre in preparations of his research papers and of his courses were crucial in the lay-out of this manuscript. Consequently, I also regard Andre as key in the writing of this monograph, and reiterate that any laud of this work must be attributed to Andre Bardhokian, Cand. Phil., more perhaps than to this author.

Third, I must acknowledge all of the predental and premedical students who have worked, and continue to work with us, as well as M.H. Ramchandani, DDS, MS, and O. Oluwadara, DDS, MS, PhD, and Professor X. Brant, DDS, Research Scholars in my research group, for their discussions, contributions, and participation in our weekly research meetings, where issues and concepts ranging from bone metabolism and cellular immunology, to systematic reviews, evidence-based clinical decision-making and comparative effectiveness analysis are routinely entertained.

Among my undergraduate students whose most salient contribution to the present endeavor I wish not to forget, I cite with praise Raisa Avezova (pre-dent), Sohrab Danaie (pre-dent), Nora Ghodousi (pre-dent), George Kossan (pre-graduate studies), Argina Kudaverdian (pre-dent), Natasha Iyer (pre-med), Nicole Mahanian (pre-dent), Linda Phi (pre-dent), and Amy Giroux (pre-dent) for their superb contribution. Without their input, this monograph could not have been crafted.

I thank Andrea Macaluso, Editorial Director, Biomedicine, Springer-US, and his dedicated staff, for his confidence and trust that this important monograph would see the light of day, despite the plethora of delays, due in part to the reasons mentioned above and in part to health issues that plagued the contributors of this work during the writing process. The unwavering support that I have received from Andrea's team throughout the many months that led to the production of this monograph is a vivid demonstration of the excellence of the director of the editorial office of Biomedicine, Springer-US, Andrea Macaluso.

It is with a sense of scientific diligence that I started this project. Today, as I complete it, I am overwhelmed by a sense of awe – again, as so often, in my scientific career – for the incredible depth and beauty, complexity in its simplicity, and simplicity in its complexity of nature, of the biological sciences, and in this particular context, of the delicate intertwined relationships between the most static of our bodily systems – bone – and the most mobile and fluid one – the immune system. I am overtaken by an overpowering feeling of admiration for the marvels of science, and for the scientific method, which leads us to its discovery. Osteoimmunology, the interrelationships between bone and immune surveillance process, is a premier example not only of the supreme magnificence of biology, but also of the excellence and power of our inductive-deductive system of scientific inquiry that permits us a glimpse and a greater glimpse yet of these biological processes and their *sequelae* in health and disease, and toward the best available evidence for treatment, and thus for continued or regained, and sustained quality of life. This sense of awe is akin perhaps to that of the poet who witnessed dawn, and wrote it all in two simple verses:

> "*M'illumino/d'immenso.*"
> (roughly rendered as *I find enlightenment in the immensity*)
> (Giuseppe Ungaretti, 1888–1970; Mattina)

Last but not least, I dedicate this work, as all of my academic endeavors to Olivia, and to Aymerica and Fredi. Moreover, this writing, as all, only and most humbly serves to further honor

> "*...la gloria di Colui che tutto move*
> *per l'universo penetra e risplende*
> *in una parte più e meno altrove...*"
> (Dante Alighieri, 1265–1321; La Divina Commedia, Paradiso, I 1–3).

Contents

1 Osteoimmunology: The Bone-Immune Crosstalk ... 1
1.1 Historical Background ... 1
1.2 Physiology ... 3
1.2.1 The Bone Talks to the Immune System ... 3
1.2.2 The Immune System Talks to the Bone ... 16
1.2.3 The Osteoblastic Niche for Immunogenesis ... 23
1.3 Implication for Stomatology ... 25
1.3.1 Facial Osteology ... 25
1.3.2 The Temporomandibular Joint ... 32
1.3.3 Immuno-Biology of Facial Bony Structures ... 35

2 Osteoimmunopathology ... 39
2.1 Inflammatory Processes ... 39
2.1.1 Osteoarthritis ... 39
2.1.2 Osteoporosis in HIV− and HIV+ Persons ... 40
2.1.3 Ankylosing Spondylitis ... 50
2.2 Bone Tumors and Necroses ... 55
2.2.1 Osteochondroma ... 55
2.2.2 Bone Malignancies ... 56
2.2.3 Osteonecrosis ... 58
2.3 Stomatological Osteoimmunopathology ... 59
2.3.1 Osteonecrosis of the Jaw ... 59
2.3.2 Temporomandibular Disorders ... 61
2.3.3 Fanconi Anemia ... 66

3 Translational Evidence-Based Interventions in Osteoimmunology ... 69
3.1 The State of the Art ... 69
3.1.1 Research Synthesis ... 69
3.1.2 Example of Existing Systematic Reviews in Osteoimmunology ... 78
3.1.3 Clinical Decision-Making ... 89

3.2 Directions for the Future ... 91
3.2.1 Translational Evidence-Based Interventions: Toward a Novel Model ... 91
3.2.2 Perfecting the Protocol for Getting the Best Available Evidence ... 92
3.2.3 Implications for Stomatology ... 99

4 Summary and Conclusion ... 101

Glossary ... 103

Notes ... 113

References ... 119

Index ... 133

Chapter 1
Osteoimmunology: The Bone-Immune Crosstalk

1.1 Historical Background

The term osteoimmunology was first used in 2000 by Professors Arron and Choi (2001). This monograph does not seek to repeat our knowledge of the fundamental inter-relationship between the bone and the immune systems. Rather, the intent of this writing is to expand our understanding of the implications and applications of this novel aspect of the basic sciences by integrating it in the emerging priorities of evidence-based clinical decision-making and comparative effectiveness research.[1]

Case in point, the recent report by the Agency for Healthcare Research and Quality (AHRQ; ahrq.gov) on the process of inflammation of joints, such as the knee joint in primary and secondary osteoarthritis, and the elucidation of the best available evidence for treatment effectiveness, with emphasis on risk-benefit and benefit-cost assessments (*vide infra*).

The purpose of this monograph is to explore the fundamentals of the bone-immune crosstalk as they pertain, for example, to the case outlined above of an inflammatory process proximal to, and detrimental to bones and joints. This monograph then discusses in greater depth the process by which the best available evidence

[1] AHRQ defines comparative effectiveness research as "...designed to inform health-care decisions by providing evidence on the effectiveness, benefits, and harms of different treatment options....", by contrast, AHRQ makes the statement that, "within the context of clinical decisions, (clinical prediction rules) pull in aspects of the history and physical exam and in an evidence based fashion estimate... or make treatment recommendations...". One can distinguish, therefore, comparative effectiveness research and analysis from clinical decision-making that is evidence-based, that is based on the best available evidence, on the grounds that the former's bottom line relates to cost-effectiveness, cost-benefit ratio, and risk-benefit ratio, and is primarily a utility-based process aimed at increasing the likelihood of success of treatment for the lowest cost and risk for a patient group with a given set of symptoms and diagnostic criteria. By contrast, evidence-based decision-making, while certainly incorporating concerns of costs, is primarily directed by the clinical needs and wants of the patients and the clinician's expertise, and involves an inductive/deductive process of logic in designing the optimal treatment of any one given patient, in an unquestionably patient-centered intent, format, and delivery (Chiappelli and Cajulis 2009; Chiappelli et al. 2009).

F. Chiappelli, *Osteoimmunopathology: Evidence-Based Perspectives from Molecular Biology to Systems Biology*, DOI 10.1007/978-1-4419-9495-0_1,

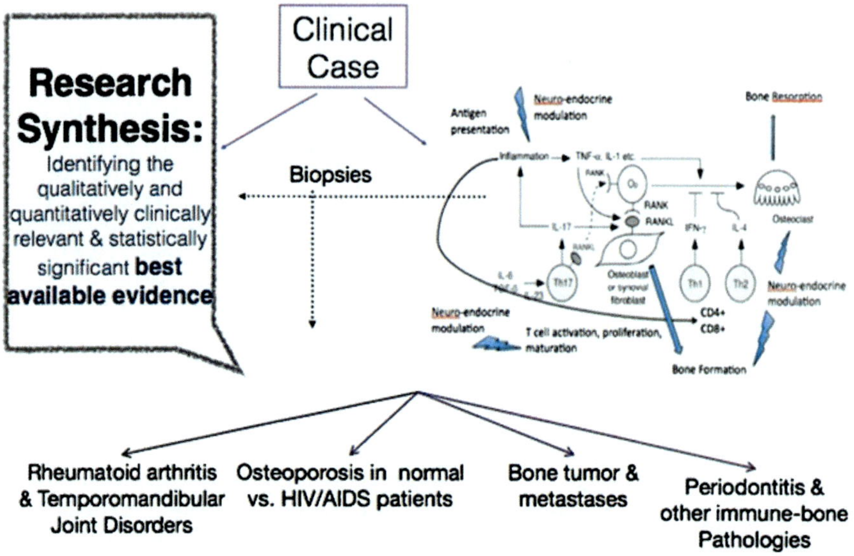

Fig. 1.1 Translational evidence-based osteoimmunology. The figure presents the fundamental paradigm of this book. The model is one that initiates with a clinical case, which, in the biological context of this work, relates to the inter-connections between the bone system and the immune system. The essential cellular and molecular components of the bone (=osteo) – immune interaction, "osteoimmunology" – are presented in the insert to the figure, and developed in the pages of this book. The figure outlines the process by which, starting from a clinical case that relates to osteoimmunology, the best available evidence for treatment efficacy and for benefit effectiveness can be obtained by means of the research synthesis design. The translational nature of this approach, with emphasis to applications and implications to a variety of osteoimmunopathologies, which range from osteoarthritic, to ospetoporosis, bone tumor immune surveillance, and other bone immunopathologies (e.g., osteonecrosis of the jaw) are noted, as they pertain to the content of the chapters that follow

is sought, analyzed, obtained, and utilized, as exemplified by the report cited above. In brief, this monograph conjoins the cutting-edge biology of osteoimmune interactions with the cutting edge science of research synthesis in an effort toward advancing the timely and critical emerging domain of translational evidence-based osteoimmunology (Fig. 1.1).

Osteoimmunology, as a scientific discipline for research and clinical practice, is merely one decade old at the time of this writing. In that span of time, a fast-growing collection of articles on specific domains of fundamental research in osteoimmunology has been published, which documents and describes the fundamental mechanisms of interactive communication between the bone and the immune systems, as well as clinical implications and applications. The generation of this cumulative knowledge was culminated by the publication of a book chapter by Professors Weitzmann and Pacifici of Emory University at Atlanta, GA (Weitzmann and Pacifici 2005a, Chap. 6), as well as a comprehensive review of the field, first in 2005 (Weitzmann and Pacifici 2005b; Lorenzo and Choi 2005), and the seminal paper by Professor Cohen (2006). In 2007, we recall as an elegant comprehensive review article coauthored by Drs. Rauner, Sipos, and Pietschmann of the Ludwig Boltzmann Institute of Aging Research, the Institute of Pathophysiology at the Center of Physiology and Pathophysiology, and the Division of Veterinary Medicine of the University of Austria (Rauner et al. 2007), as well as an important update of the state of research in the field produced by Professor Jean-Pierre David (2007). Work continued to multiply worldwide in this domain of science, which led to the reviews by Drs. Julian Quinn and Hasnawati Saleh from Victoria Australia (cf., Lorenzo et al. 2008; Quinn and Saleh 2009), Professor Takayanagi at the University of Tokyo, Japan (Takayanagi 2009), Drs. Gallois, Mazzorana, Vacher, and Jurdic from the University of Lyon, France (Gallois et al. 2009), and Dr. Xing at the University of Rochester, NY (Xing 2009), to cite only a few salient excellent overviews of the field in the recent past (cf., Nakashima and Takayanagi 2009).

The first International Conference on Osteoimmunology: Interactions of the Immune and Skeletal Systems was held in Crete, Greece (May 28-June 2, 2006), and the proceedings, which provide an important overview and introduction to the basics of the field were published by Springer-Verlag (Choi 2007). The second International Conference on Osteoimmunology was held in Rhodes, Greece (June 8–13, 2008). Professor Yongwon Choi of the University of Pennsylvania again edited the proceedings of the conference (Choi 2009). The third International Conference on Osteoimmunology: Interactions of Immune and Skeletal Systems just met, at the time of this writing, at the Nomikos Conference Center in Fira, Santorini, Greece (June 20–25, 2010). We should look forward to the proceedings of that meeting shortly.

1.2 Physiology

1.2.1 The Bone Talks to the Immune System

The most prominent functions of bone are the protection of internal organs and the support of body structures. Beyond those functions, bone additionally serves as an attachment site for muscles allowing locomotion and as an appropriate cavity for hematopoiesis in bone marrow. As a reservoir for inorganic ions (e.g., calcium), bone is responsible for the maintenance of calcium homeostasis and is able to rapidly mobilize mineral stores on metabolic demand. Bone is a connective tissue composed of cells and extracellular matrix, the latter being further subdivided into an inorganic, and an organic component, which is mineralized. Its main constituent is type I collagen (~95%), and in minor amounts other types of collagens, noncollagenous proteins, and proteoglycans. The inorganic matrix of bone predominantly contains calcium and phosphorus in the form of hydroxyapatite crystals ($[3Ca_3(PO_4)_2](OH)_2$) deposited and amassed into the collagenous matrix. The interdigitation of the two matrices confers the characteristic rigidity and strength to the bony skeleton, while preserving some degree of elasticity. Rather than an inert and static material, bone, the major constituent of the skeleton in all vertebrates, is a highly organized living tissue whose metabolism is intimately intertwined with, cross-regulates, and is modulated by several of the major physiological systems.

The immune system, and the cells and factors that constitute it, provides an essential and integral part of the underlying metabolism of bone. These processes are reviewed in the following sections. In brief:

- *Osteo-nervous interactions* pertain to the observation that bone is richly innervated by both autonomic and sensory neurons, serve sensory and regulatory functions, and mediate bone cell and immune cell activities directly (Warden et al. 2005).
- *Osteo-endocrine interactions* refer to our understanding of the extent to which hormones (e.g., adrenocorticotropin hormone [ACTH]; Isales et al. 2010; parathyroid hormone [PTH] and calcitonin; Carter and Schipani 2006) regulate bone mass by directly influencing the metabolism of bone-regenerating (i.e., osteoblasts) and bone-resorbing cell populations (i.e., osteoclasts) (*vide infra*), modulating either the collagen mass (i.e., ACTH) or the calcification content (i.e., PTH, calcitonin). PTH raises blood calcium levels by stimulating bone resorption. Calcitonin reduces blood calcium by suppressing bone resorption and increasing osteoid calcification. The osteoid is the matrix secreted by osteoblasts and osteocytes prior to mineralization. Calcitonin acts by directly inhibiting osteoclast activity via the calcitonin receptor. Calcitonin receptors have been identified on the surface of osteoclasts. Calcitonin directly induces inhibition of osteoclastic bone resorption by affecting actin cytoskeleton, which is needed for the osteoclastic activity. In brief, the peptide hormone PTH is one of the most important regulators of calcium ion homeostasis (Kronenberg et al. 1993; Potts et al. 1997). In response to low blood calcium levels, PTH is secreted into the circulation and acts on kidney, bone, and intestine to maintain blood calcium concentrations. In bone, PTH upregulates the production of the pro-inflammatory cytokine, interleukin (IL)-6 and of the Receptor Activator for Nuclear Factor κ B Ligand (RANKL; also known as TRANCE: Tumor Necrosis Factor [TNF]-related activation-induced cytokine, OPGL: osteoprotegerin ligand, and ODF: osteoclast differentiation factor) (*vide infra*) by osteoblasts, thereby facilitating the differentiation, activation, and survival of osteoclasts (Huang et al. 1998; Dai et al. 2006). Consequently, PTH, as well as PTH-related protein (PTH-rP), promotes bone resorption and de facto the release of calcium (Pollock et al. 1996; Onyia et al. 1997). The active hormonal form of vitamin D (aka, calcitriol, 1,25-dihydroxycholecalciferol, 1,25-dihydroxyvitamin D_3, VitD), is essential for the development and maintenance of the mineralized skeleton (Dardenne et al. 2001; Panda et al. 2004). Osteoblast number, bone formation, and bone volume increase serum alkaline phosphatase levels, and are associated with a decreased production of RANKL, but an enhanced production of the osteoclastic cytokine osteoprotegerin (OPG; also known as OCIF: osteoclastogenesis inhibitory factor) (*vide infra*) (Kitazawa et al. 2003). Regarding bone homeostasis,

estrogen and androgens are the most intensively investigated sex steroids and, in contrast to PTH and Vit D, enhance bone formation and inhibit bone resorption (Hofbauer et al. 1999; Khosla et al. 2002; Leder et al. 2003). Estrogen as well as testosterone deficiency inevitably lead to an increased rate of bone turnover (Jilka et al. 1998), as well as simultaneous increases in osteoclastic precursors and early osteoblastic precursors. Estrogen deficiency results in a net loss of bone as a consequence of an increased production of RANKL and a decreased production of OPG by osteoblastic cells, as well as the enhancement of the secretion of pro-inflammatory and pro-resorptive cytokines by lymphocytes such as IL-1, IL-6, and TNF-α (Pacifici et al. 1991; Jilka et al. 1995; Eghbali-Fatourechi et al. 2003). The bone-protective effect of estrogen is mediated in large part by transforming growth factor (TGF)-β, which induces apoptosis of osteoclasts (Oursler et al. 1991; Hughes et al. 1996; Fox and Lovibond 2005). From the viewpoint of research synthesis for evidence-based and comparative effectiveness analysis, as discussed in greater details in Chap. 3, it is important to note at this juncture that in at least two randomized controlled trials from the Women's Health Initiative, hormone replacement therapy (HRT) was shown to decrease the incidence of major osteoporotic fractures. Serious undesirable side effects such as cardiovascular disease and cancer have occurred and therefore other medications are used nowadays in the treatment of osteoporosis. Raloxifene is a selective estrogen receptor modulator (SERM) and is approved for the treatment of osteoporosis. Like estrogen, SERMs are known to mediate their effects through the estrogen receptor. While estrogen binds equally strongly to α- and β-receptors, raloxifene preferentially binds to the α-receptor (Women's Health Initiative 2002, 2004). From the perspective of translational clinically relevant complex systematic reviews (Chiappelli et al. 2010a, b), this specificity to a certain estrogen receptor will have implications in the evidence-based revision of clinical practice guidelines because it points to a fundamental biological process that allows a higher affinity to bone, and therefore the risk of side effects of SERMs are less pronounced than those of HRT (Riggs and Hartmann 2003). Similarly, as bone metabolism is inter-dependent with the endocrine system, and it is regulated and modulated by the central and peripheral nervous systems (cf., neuroendocrinology), it is not surprising to find that neuropeptides (e.g., neuropeptide Y, NpY; gastrointestinal peptides, etc.) modulate osteoblasts and osteoclasts, and have direct regulatory effects upon bone metabolic activity (Lee and Herzog 2009; Sousa et al. 2009; Wong et al. 2010).

- *Osteoimmune interactions* refer to the involvement of immune cells and the factors they produce (e.g., cytokines, *vide infra*) as they modulate bone metabolism, as is discussed in greater depth in the remainder of this book. Suffice to state at this juncture that since osteoclasts are derived from the monocyte/macrophage (=myeloid) lineage, the macrophage-osteoclast interaction is of prime importance in bone metabolic regulation. Certainly, evidence is mounting in support of osteoclast interaction with T cells, B cells, and dendritic cells (Takayanagi 2010). Moreover, the emergence and replenishment of the immune system throughout the life span occurs as hematopoietic precursors develop in the bone marrow, a process that is finely regulated within the neural-immune-hematopoietic axis by neuropeptides of the tachykinins family (e.g., substance P, neurokinin [NK]-A) (Greco et al. 2004; Murthy et al. 2007). Evidently, these physiological facets interplay with the two natural kinds of mature bone: cortical compact bone and medullary spongy bone.

- *Compact and dense cortical bone* typically composes most of the thickness of long bone shaft, the diaphysis (Gr: dia, through; π, phusis, structure, formation) is made of compact bone. Micro-anatomical studies reveal that compact cortical bone is composed of cylindrical units of bone structure, the osteons with a dimension of circa 0.2 mm in diameter. Osteons together form the Harvesian system, named after the British physician and anatomist Clopton Havers (1657–1702). Each osteon consists of concentric lamellae of bone matrix, mainly collagen fibers, surrounding the central Harvesian canal. These micro-anatomical structures form canal-like structures, which Havers himself termed Canals of Havers in his *Osteologia Nova*, (Havers, 1691). Blood vessels and nerves course the canals, and thusly penetrate the bone tissue. The long axis of osteon is usually parallel to the long axis of the bone, but the collagen fibers in the different lamellae of the osteon are oriented at different angles and provide increased strength and elasticity of the osteon units. Volkmann's canals, named after the German physiologist and anatomist Alfred Wilhelm Volkmann (1800–1877), constitute an alternative series of channels that provide a route for blood vessels and nerves to reach the principal osteonal canal, and that link together into a network that inter-connects the Harvesian canals across different osteons. This micro-anatomical network ensures the physiological homogeneity of bone metabolism, and the continuity of circulatory and nervous supply to the bone marrow cavity. Within that histological contextual milieu, osteoblasts develop and mature into osteocytes, each living within its own micro-environmental *lacuna*. Osteocytes make contact with the cytoplasmic processes of their counterparts across *lacunae* via a network of even smaller canals, the *canaliculi*, which together form a structure that facilitates the exchange of nutrients and metabolic waste among osteocytes and osteon units. In brief, nutrients and other substances, including hormones, neuropeptides and cytokines, pass from blood vessels into the Harvesian canal, through an increasingly finer *reseau* of channels, eventually to distant osteocytes. Inner and outer circumferential *lamellae* run the length of the shaft located at the inner and outer surface of the long bone. Osteocytes sit in the *lacunae* between the *lamellae*, and as such can be involved in both bone deposition (i.e., osteoblasts and mineralization) and bone resorption (i.e., osteoclasts and bone degradation), both of which are modulated (i.e., increased under certain conditions, and decreased in other cases) by products associated with the cell-mediated immune system (e.g., cytokines, cytokine receptors, membrane-bound and soluble clusters of differentiations [CD's; cf., Notes], and the like). Interstitial lamellae form as remnants of previous osteons, and reflect the process of constant bone modeling, resorption, and remodeling (i.e., bone metabolism), which is orchestrated by the interaction of bone cells with cells of immune system within and about these micro-anatomical structures of osteons, canals, *caniculli*, *lamellae,* and *lacunae.*

- The *spongy, cancellous, trabecular, medullary bone* is composed of similar cells and matrix structure, and in that resembles compact bone. However, micro-anatomical examination reveals a distinct lamellar structure of collagen in spongy bone, which is typically not arranged concentrically around a central canal, but rather as *lamellale* that run parallel to one another. Consequently, spongy bone is composed of bone spicules, the *trabeculae*, which are endowed of varying shapes and sizes. The space between the spicules is filled with the bone marrow, the flexible tissue found in the hollow interior of bones. That space is rich in nerve endings, blood vessels, and capillaries. This micro-anatomical histological "geographical" distribution is critical in the formation of niches, where cells of the bone system – the osteoprogenitor cells at the external and internal surface of the bone – and cells of the immune system – the hematopoietic progenitor cells – develop, interact, communicate, and cross-feed in a fundamental osteoimmune concert of epigenetic events (*vide infra*).
- The *bone marrow* is the underlying contextual framework wherein the interaction between the bone and the immune system commences. The red marrow consists mainly of hematopoietic tissue, whereas the yellow marrow, which constitutes the stroma of the bone marrow, contains principally adipocytes.
- *Compact and spongy bone interact*: Small congregates of spongy bone result that are facing the marrow cavity, the large medullary cavity. The two expanded ends of the long bones, the anepiphyses (Gr: ana, end-piece; phusis, structure, formation), consist mostly of spongy bone covered with a thin shell of compact bone. The calvarium and the sternum are two flat bones made of two layers of relatively thick compact bone, and an intervening layer of spongy bone.
- Bone is lined externally by a dense connective tissue, the *periosteum*, composed histologically of an outer fibrous layer, rich in fibroblasts, and an inner osteogenic layer, rich in periosteal and endosteal cells that give rise to the osteoblast population. The inner side of bones is lined by the endosteum. There is no periosteum lining at the joints of long bones. In remodeling bone, periosteal cells work in concert with bone modeling factors, including cytokines and other immune factors, to increase bone width and length. In nonremodeling bone by contrast, that surface is lined by a layer of flat bone lining cells, which appear quiescent and incapable of osteogenesis. It is believed that these cells still may function as nutritional support for osteocytes that are embedded in the underlying bone matrix.
- *Endosteal cells* line the marrow cavities in compact bone, and the spicules of spongy bone. In a manner akin to the periosteal cells, endosteal cells act as osteoprogenitor cells, and are modulated by immune factors to give rise to osteoblasts. Indeed, research has established that two steps occur in bone remodeling, which involve the endosteal cell layer: resorption of a volume of bone by osteoclasts is followed sequentially by the formation of a comparable volume by osteoblasts. However, the regulatory processes that initiate, sustain, and terminate this sequence are intertwined, and intimately regulated by signals, which travel along the osteocyte canalicular system to endosteal lining cells, and

they entertain a complex molecular cross-talk that involves precursors, mature cells, cells of the immune system, and products of both the resorbed matrix and cellular immune products that titrate each other, and modulate each other in a concerted and finely orchestrated cross-system, multicellular remodeling machinery toward the end-result of either removing or forming a net volume of bone, osteoprecursors, or hematopoietic precursors (Martin and Seeman 2008; Matsuo and Irie 2008). Bidirectional signaling and interaction is likely to occur and to continue among osteoblasts, osteoclasts, and endosteal lining cells throughout the lifespan, during which time bone metabolism is both sustained and modulated by immune factors, and results in osteoblastic bone formation with mineralization of bone matrix, and osteoclastic bone resorption with apoptosis and demineralization.

- The *bone remodeling cycle* involves a complex series of sequential steps that are highly regulated. The "activation" phase of remodeling is dependent on the effects of local and systemic factors on mesenchymal cells of the osteoblastic lineage. These cells interact with hematopoietic precursors to form osteoclasts in the "resorption" phase. In the later "reversal" phase, mononuclear cells appear on the bone surface, complete the resorption process, and produce the signals that initiate formation. In successive waves, mesenchymal cells newly derived from the adult stem cell pool differentiate into functional osteoblasts, which lay down matrix in the next "formation" phase. Case in point, new research demonstrates that vascular endothelial cells can transform into multipotent stem-like cells, which by all account resemble mesenchymal stem cells, by means of an activin-like kinase-2 (ALK2) receptor–dependent mechanism: activation of the ALK2 pathway (cf., Notes) in endothelial cells by means of ligands such as with the ligands transforming growth factor-β2 (TGF-β2) or bone morphogenetic protein-4 (BMP4), led to a distinct endothelial-to-mesenchymal transition and acquisition of a stem cell-like phenotype (Medici et al. 2010) (Fig. 1.2).
- *Osteoblasts* are immature bone cells, specifically mononucleate bone-forming cells that descend from osteoprogenitor cells. Osteoblasts derive from a common family branch that arises from mesenchyme, a type of loose connective tissue, derived from the three embryonic germ layers (i.e., endoderm, mesoderm, ectoderm). The mesoderm is the proper germ layer that gives rise to the skeleton and the hematopoietic system. The mesenchymal prominent ground substance matrix contains a loose aggregate of reticular fibrils and unspecialized cells, which are capable of developing into bone, cartilage, lymphoid organs, and the circulatory system (Strum et al. 2007). Core binding factor 1 (Cbfa1) the runt-related transcription factor 2 (runx2), and the downstream factor osterix (Osx) are critical transcription factors for lineage commitment of stem cells toward osteoblast differentiation (Ducy et al. 1997; Caetano-Lopes et al. 2007). Although a relatively rare event, it is not excluded that, given the appropriate microenvironment, osteocytes and osteoblasts can revert back to earlier stages of their development. Note that most current research has established that endothelial cell-derived mesenchymal stem cells may be an efficient

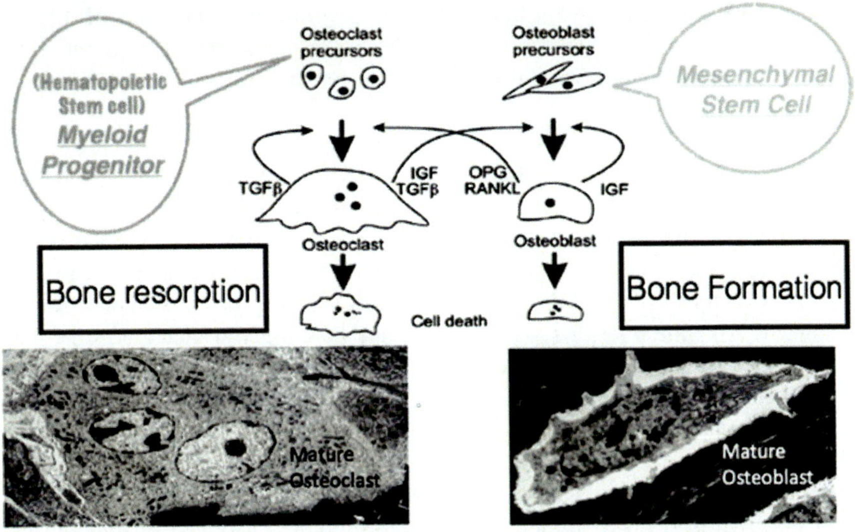

Fig. 1.2 Bone metabolism. The figure shows a composite of electron micrographs of, respectively, a mature human osteoclast, and a mature human osteoblast. Characteristically, the relative myeloid-related complexity of the cytoplasmic compartment of the former contrasts with the relative smoothness of the mesenchyme-derived osteoblast cytoplasm. Integrated in the figure is a diagrammatic representation of the maturation process of both populations from their respective myeloid osteoclat precursor and mesenchyme osteoblast precursor. The central aspect of the figure represents the fundamental core of bone metabolism. The balance between bone formation by osteoblasts and bone resorption by osteoclasts is regulated by immunological factors, such as tumor growth factor (TGF)β, insulin-like growth factor (IGF), osteoprotegerun (OPG), which blocks receptor activator of NF-kB ligand (RANKL) signaling in osteoclasts. It is important to note that among the families of regulatory proteins for bone metabolism, a large body of research has emerged in the most recent decade on the role of the family of Wnt's, which are secreted cysteine-rich glycoproteins that locally activate receptor-mediated signaling pathways, in part because mutations of Wnt's that lead to either gain or loss of function of the Wnt coreceptor lipoprotein receptor-related protein 5 (Lrp5) induce significant changes in bone mass (e.g., high bone mass) or osteoporosis, respectively. It is now clear, moreover that the "canonical" (i.e., b-catenin-dependent; b-catenin is an intracellular anchoring protein involved in cell adhesion through E-cadherin) Wnt signaling pathway regulates the lineage progression for the maturation of osteoblasts, and represses osteoclastogenesis by increasing osteoprotegrin (OPG) expression, thus altering the OPG/RANKL ratio. Adapted from: ebi.ac.uk/biomodels-main/static-pagesdo?page=ModelMonth%252FOctober2007%252FBIOMD0000000148_MM; endotext.org/parathyroid/parathyroid1/parathyroid1.html

source of cells for regeneration of bone and cartilage (Horwitz 2010). Indeed, preliminary results indicate that endothelial cell–derived mesenchymal stem cells may actually be superior as a cell source for bone and cartilage cell therapy for patients requiring regenerative medicine interventions (Medici et al. 2010).

- Osteoblasts are located on the surface of *osteoid* seams and make a protein mixture known as the osteoid, which mineralizes to become bone. The osteoid seam is a narrow region of newly formed organic matrix, not yet mineralized, located on the surface of a bone. Osteoid is primarily composed of Type I collagen produced by the osteoblasts, which also manufacture hormones and prostaglandins that mediate bone metabolism for the formation of the immediately surrounding matrix.

- The principal *proteomic signature of osteoblasts* is the production and expression of alkaline phosphatase (EC 3.1.3.1), a key dephosphorylating enzyme that contributes to the accumulation of calcium and phosphate into the vesicles generated during the formation of the matrix. Four principal isozymes of alkaline phosphatase have been recognized. The alkaline phosphatase, tissue-nonspecific isozyme is found in bone, and is encoded by the ALPL gene, located on chromosome 1. Alkaline phosphatase is a membrane-bound glycosylated enzyme that is not expressed in any particular tissue and is, therefore, referred to as the tissue-nonspecific form of the enzyme. When missing, the disorder known as hypophosphatasia arises, which manifests as hypercalcemia and skeletal defects (Swallow et al. 1998).
- Osteoblasts also produce bone sialoprotein, a 60–80-kDa small integrin-binding ligand, N-linked glycoproteins (SIBLING), protein constituent of mineralized tissues such as bone, dentin, cementum and calcified cartilage, osteocalcin (also known as BGLAP: bone γ-carboxyglutamic acid-containing protein), and osteopontin (OPN; also known as also known as BSP-1 or BNSP: bone sialoprotein I, ETA-1: early T-lymphocyte activation, or SPP1: secreted phosphoprotein 1). OPN is a remarkable member of the SIBLING family that is particularly interesting to osteoimmunologists because, while produced by osteoblasts and critical of osteoblastic function, it is also expressed by a variety of immune cells, including macrophages, neutrophils, dendritic cells, and T and B cells. In the immune system, OPN is reported to be endowed with chemotactic properties that promote cell recruitment to inflammatory sites, adhesion properties by means of binding to several integrin receptors, a function that favors cell attachment and contributes to promoting cellular immune activation of T cells and cytokine production, as well as cell survival and regulation of apoptosis. The SIBLING protein family is a group of noncollagenous proteins whose members (e.g., OPN, BSP) appear at distinct phases of development, which suggest substantial differences in the distribution of the SIBLING proteins between organic and inorganic phases, and divergent molecular and epigenetic regulatory roles in osteogenesis and immune regulation of osteogenesis (Weber and Cantor 1996; Huang et al. 2008; Wang and Denhardt 2008; Sun et al. 2010) (cf., notes).
- In brief, therefore, and by their very nature, osteoblasts occupy a central place in the *osteoimmune dialog*: they support hematopoiesis, and when completely surrounded by matrix, osteoblasts become osteocytes, which can both secrete and resorb matrix, and thus are responsible for maintaining the bone matrix itself. In that respect, they act as mechano-sensory receptors of the osteoimmune entity, capable of regulating the bone's response to stress and mechanical load. Osteocytes are typically poor in organelles, indicating other primary functions than matrix synthesis and mineralization. Further maturation of osteocytes alters their morphology into cells projecting dendritic processes. These channel-like structures enable osteocytes to communicate with each other, may, among other

functions, act as mechanosensors to permit bone to react to environmental challenges. Loaded vesicles eventually rupture, and increase local concentration of minerals that contributes to initiate the mineralization process. Osteoblasts and osteocytes synthesize the collagen-rich organic matrix that provides the optimal conditions for matrix mineralization by secreting numerous bone matrix proteins and matrix metalloproteinases (MMP). They eventually engage in programmed cell death, apoptosis (cf., notes), which leads to increased secretion of osteoclastogenic cytokines that favor bone resorption (Gu et al. 2005; Kogianni et al. 2006). It is now clear that apoptosis is not the only regulated cell death program involved in the concerted modulation of tissue homeostasis and the removal of unwanted cells in biological organisms. Other cell death modalities, and their cross-talk, require increased understanding (Zhivotovsky and Orrenius 2010), particularly as they pertain to immune-mediated bone degenerative diseases.

- By contrast, *osteoclasts* are responsible for remodeling of bone to reduce its volume – that is, of bone resorption. These large multinucleated cells – tissue-specific giant polykaryons – arise from monocyte common progenitors, and can actually arise from monocyte/macrophages directly, and are thus not related to the same family lineage of bone forming cells as osteoblasts. Osteoclasts are multinucleated, giant cells of hematopoietic origin formed by the fusion of mononuclear preosteoclasts derived from myeloid cells. Fusion-mediated giant cell formation is critical for osteoclast maturation: without it, bone resorption is inefficient. The d2 isoform of vacuolar (H+) ATPase (v-ATPase) V0 domain (Atp6V0d2) is one of the principal regulators of osteoclast fusion and bone formation. Similar to the myeloid family, osteoclasts are endowed with phagocytic properties, share several families of plasma membrane receptors with certain immune cell populations, and function in a manner similar to their mature myeloid equivalent. They are found on bone surfaces in what are called Howship's *lacunae*, named after John Howship (British surgeon, 1781–1841). These structures are bone resorption pits that result following the breakdown of the bone surface and consequential erosion of the bone by the osteoclasic enzymes, including lysosomes, organic acids, and hydrolytic enzymes. The osteoclastic layer that contacts the bone is divided into the microvillus structure, a ruffled border rich in plasma membrane folding, and a ring-like perimeter of cytoplasm, termed the clear zone, which marks the area of bone in the process of being resorbed.
- The *metabolic function of osteoclastic structures*, which we recall from our introductory statements, is modulated by the neuroendocrine system, such that an increase in PTH levels, and a decrease in calcitonin is associated with an increased number and functionality of osteoclasts. PTH is secreted by the parathyroid glands as an 84 amino acid polypeptide, which acts to increase the serum concentration of calcium (Ca^{2+}). Calcitonin is a 32-amino acid linear polypeptide hormone produced primarily by the parafollicular cells, the C-cells of the thyroid, which are located adjacent to the thyroid follicles and reside in the connective tissue. They appear large and with a characteristic pale stain, compared with the follicular cells or colloid. Calcitonin acts to counter PTH, and to reduce blood calcium.

- *Osteoclastic factors* shared with the hematopoietic cell lineage briefly noted in our introductory remarks include:
  - *OPG* (i.e., osteoprotegerin), the OCIF is a cytokine-like 401 amino acid peptide found either as a 60-kDa monomer or 120-kDa dimer linked by disulfide bonds, first reported as a protein that exposed an osteopetrotic phenotype when overexpressed in transgenic mice, which was secreted by preosteoblasts/stromal cells, which could inhibit osteoclast development and activation (hence the name attributed to its bone-protective effects: osteoprotegerin) (Simonet et al. 1997). OPG belongs to the TNF receptor superfamily, but lacks the transmembrane and cytoplasmic domain. OPG is a cytokine expressed in a variety of tissues (e.g., lung, heart, kidney, liver, stomach, intestine, brain, spinal cord, thyroid gland, smooth muscle tissue, and, in addition to osteoclasts, certain immune cells – in fact, the expression of OPG by dendritic cells, a population of antigen-presenting cells, was shown to increase with immune maturation, to be stringently dependent upon NF-κB signaling, and to act in concert with regulatory processes of immune responses in lymphoid tissues (Schoppet et al. 2007)). Nonetheless, the principal osteoimmune role of OPG has been assigned to bone protection by impairing the function and maturation of osteoclasts. This is obtained by OPG binding to RANKL on osteoblast/stromal cells, thus blocking the RANKL–RANK ligand interaction between osteoblast/stromal cells and osteoclast precursors, thereby blunting the differentiation of the osteoclast precursors into mature osteoclasts (Khosla 2001; Boyce and Xing 2007). OPG also promotes cell survival, in certain physiological situations, by inhibiting TNF-related apoptosis-inducing ligand (TRAIL)-induced apoptosis (Reid and Holen 2009).
  - The *production of OPG is subject to neuroendocrine regulation*. Specifically, production of OPG is stimulated in vivo by the female sex steroid estrogen, and accounts for the biochemical mechanisms by which estrogen is the predominant sex hormone that slows bone loss. As the production of estrogen becomes irregular, and decreases with age (i.e., menopause), the likelihood of osteoclast-mediated bone resorption, and consequentially osteoporosis (*vide infra*), increases. One treatment modality for osteoporosis includes the drug strontium ranelate, which both increases deposition of new bone osteoblasts and reduces the resorption of bone by osteoclasts. In fact, fundamental research has established that strontium ranelate acts by means of activation of the calcium- or other cation-sensing receptor, by increasing the expression of OPG, coupled with decreasing RANKL expression by the osteoblasts (Hamdy 2009).
  - *RANKL* is the ligand for the receptor activator for Nuclear Factor |B, RANKL is also known as TRANCE: TNF-related activation-induced cytokine, OPGL: ligand for OPG, and ODF. The expression of OPG and RANKL is modulated by various endocrine, immune, inflammatory, and neuroendocrine-immune modulators, including PTH, estrogens, glucocorticoids IL-6, IL-8, IL-11, INF-γ, TGF-β, prostaglandin (PG)-E2, bone morphogenetic protein (BMP)-2, VitD, and many others (Khosla 2001). The extensive distribution of RANKL throughout the body attests for its multiple functions, among which the

most important – at least in our context presently – is the induction and regulation of osteoclastogenesis. RANKL is a surface-bound molecule on certain immune cells, and on osteoblasts that serves to activate osteoclasts. Overproduction of RANKL has been implicated in a variety of degenerative bone diseases (e.g., rheumatoid arthritis, psoriatic arthritis). Targeted silencing of the related gene in a murine experimental system leads to a lack of osteoclasts and associated osteopetrosis. RANKL, which is also expressed by memory T helper cells, plays a role in dendritic cell survival and maturation, and modulates the cellular immune system by contributing to the regulation of T cell activation, proliferation, and maturation. Activated T cells induce expression of the RANKL gene, which, in the osteoimmune context, contributes to osteoclastogenesis and increase bone resorption and loss. The human RANKL gene has been localized to chromosome 13q14 and encodes for three isoforms: RANKL1 and RANKL2 are type II transmembrane proteins, and RANKL2 also possesses a shorter intracellular domain. RANKL3 is a soluble protein, partially produced by the cleavage of the membrane-bound form by TACE (TNF (converting enzyme, a metalloprotease; cf. notes)).

- The *RANKL proteomic signature* pathway includes the activation of RANK following binding of its ligand (RANKL), and leading to activation of AKT/PKB. This is an enzyme complex endowed with antiapoptotic serine/threonine kinase, protein kinase B (PKB) activity, which is also called AKT.[2] As of now, three genes have been identified in the Akt family: Akt1, 2, and 3, which code for enzymes that are members of the serine/threonine-specific protein kinase family (EC 2.7.11.1). Akt1 is involved in cellular survival and it inhibits the apoptotic processes. Akt1 also induces protein synthesis pathways, and, in general, tissue growth and development. Akt2 is an important signaling molecule in the insulin signaling pathway, and is required to induce glucose transport. Akt3 is predominantly expressed in brain, and mediates the regulation of neurogenesis. The Akt enzyme family possesses a protein Pleckstrin Homology (PH) domain, which allows it to bind to phosphoinositides with high affinity, thus effecting the kinase signaling pathway. In this manner, once correctly positioned in the membrane via binding of PIP3, Akt is phosphorylated by its activating kinases (i.e., phosphoinositide dependent kinase 1 (PDPK1) at threonine 308, and mammalian target of rapamycin complex 2 (mTORC2), a phosphatidylinositol 3-kinase-related kinase, at serine 473). The mTOR phosphorylation events stimulate the subsequent phosphorylation of Akt by PDK1. Phosphatidylinositol-3 kinase (PI3K)-dependent Akt activation can be regulated through the tumor suppressor phosphatase and tensin homolog (PTEN). Akt may also be activated in a PI 3-kinase-independent, and cAMP-dependent protein kinase A (PKA)-dependent activity. That is to say, convergent

[2] The Akt nomenclature is rather unusual, if not enigmatic as it is not descriptive of a function per se: "Ak" was a temporary classification name of a murine strain with spontaneous thymic lymphomas, and "t" was simply meant to refer to "transforming" (Staal et al. 1977).

pathways ensure activation of AKT/PKB, because, once activated, this enzyme complex can go on to activate or to deactivate its myriad substrates via its kinase activity, and to regulate cellular survival and metabolism by binding and regulating a plethora of downstream effectors. The AKT/PKB pathway engaged by RANKL is critical to the regulation of several cellular processes in bone and in immune cells through a multipronged proteomic profile.

- *RANK*, the Receptor Activator of Nuclear Factor κB, is a member of the TNF receptor family, which also includes CD120 (TNFα receptor), CD40, Fas, and CD34. The family of proteins, which consists to date of 6 members (TRAF1-6), is endowed with complex functional properties in that they converge in regulating inflammation, antiviral responses, and apoptosis by mediating transmembrane signaling not only for the members of the TNF receptor family, but also for the members of the Toll/IL-1 family. TRAF proteins interact with several protein kinases including IRAK1/IRAK, SRC, and PKCz, and thus establish a critical regulatory link between distinct signaling pathways. Thus, and in the specific context of the present discussion, RANK signaling, initiated by RANKL, can be transferred to a parallel cascade via TRAF-6, which, upon activation engages signaling pathways leading to the activation of the transcription factors NFAT, NFkB, the MAP kinase mediators jun, fos, and p38 as well as the down-stream targets of Akt AFX/FOXO4. The two most investigated pathways are the activation of the transcription factors NF-κB and AP-1 (activated protein 1). Targeted disruptions of the p50/p52 component of NF-κB and the c-fos component of AP-1 result in impaired osteoclastogenesis and an osteopetrotic phenotype (Kobayashi et al. 2001; Ye et al. 2002). Together, these signals (cf., notes) contribute to osteoclast differentiation, activation, and survival. Indeed, RANK is a type I membrane protein expressed by osteoclasts, as well as by dendritic cells, and involved in bone resorption and the facilitation of immune signaling. RANK-deficient mice show similar phenotypes to those of RANKL knock-out mice, including osteopetrosis and missing lymph nodes (Dougall et al. 1999).

Furthermore, we must note in the osteoimmune spectrum:

- *M-CSF*: the macrophage-colony-stimulating factor is principally a cytokine that regulates hemopoietic stem cell differentiation into macrophages and related cell types, hence, its role osteoclast differentiation. M-CSF binds to the Colony stimulating factor 1 receptor, (CSF1R; aka, macrophage colony-stimulating factor receptor, M-CFSR, CD115), a tyrosine kinase transmembrane receptor of the CSF1/platelet-derived growth factor (PDGF) receptor family.
- *Osteocalcin* is the BGLAP, a noncollagenous protein of bone and dentin, encoded by the BGLAP gene on chromosome 1. As a hormone, osteocalcin induces the beta cells of the pancreas to increase release of insulin, and directs adipocytes to release the hormone adiponectin, which increases sensitivity to insulin (Lee et al. 2008). Indeed, it is now evident that, through the uncarboxylated form of the osteoblast-derived factor osteocalcin, bone regulates glucose metabolism and fat mass, and thus has a central homeostatic role, not only for bone metabolism. Specifically, research has now established a putative working

model, which stipulates that osteoblastic function is negatively regulated by the adipocyte-produced hormone leptin. When bound to its receptor in the central nervous system (CNS), leptin exerts a stimulation of the sympathetic nervous system. Leptin deficiency leads to increased osteoblast activity and increased bone mass. By contrast, expression of Esp gene by osteoblasts regulates glucose homeostasis and adiposity by regulating osteocalcin, which mediates both pancreatic insulin and adipocytic adiponectin production for the overall modulate of energy metabolism (Wolf 2008).

- *COX2*: cyclo-oxygenase 2, the prostaglandin-endoperoxide synthase 2, acts both as dioxygenase and as a peroxidase. It is key in the biosynthesis of the prostanoids; that is, prostaglandins, prostacyclin, and thromboxanes, which contribute to the inflammatory and mitogenesis cascades. It presents as two isoenzymes: a constitutive form, COX-1, and its inducible counter-part, COX-2, which, in humans, is expressed in a limited number of cell types and regulated by certain specific stimulatory events. Interestingly, the product of the COX-2 peroxidase function on arachidonic acid, prostaglandin H2 (PGH2) is a central precursor to this family of pro-inflammatory molecules: PGH2 is converted by prostaglandin E2 synthase into PGE2, by prostaglandin D2 synthase into prostaglandin D2, by thromboxane-A synthase into thromboxane A2, and by prostacyclin synthase to create prostacyclin. Among other functions, and specifically to the focus of this writing, PGE2 contributes to the fever response, and stimulates osteoblasts to release factors to bone resorption by osteoclasts (Watkins et al. 2003; Li et al. 2006). By contrast, PGD2 recruits TH2 cells, contributes to the development and the exacerbation of allergic diseases, and, at least with respect to body temperature, acts to oppose PGE2. Note: PGH2 is also produced by COX-1. Aspirin irreversibly inhibits COX-1, thus preventing the formation of PGH2, and therefore thromboxane A2, thus blocking its regulation of the activation of new platelets and increased platelet aggregation. PGH2 contrasts the effects of prostacyclin (PGI2), which chiefly prevents formation of the platelet plug involved in primary hemostasis (i.e., blood clot formation).
- *FLAP*: The 5-lipoxygenase-activating protein (FLAP) activates the 5-lipoxygenase enzyme (aka, arachidonate 5-lipoxygenase, 5-lipoxygenase, 5-LO, Alox5), a member of the lipoxygenase family, which transforms essential fatty acids into leukotrienes. Again, two major groups of products, result that have counter-balancing effects upon the inflammatory process: arachidonic acid yields the 4-series cysteinyl leukotrienes (LTB4, LTC4, LTD4, LTE4) that are generally pro-inflammatory and make up the slow-reacting substance of anaphylaxis (SRS-A). Leukotrienes; eicosapentaenoic acid yields the 5-series Leukotrienes (LTB5, LTC5, LTD5, LTE5) that generally favor an antiinflammatory response. Whereas the role of leukotrienes, and therefore of FLAP, in immune regulation is evident, their significance to bone metabolism is unclear at this time. Nonetheless, emerging evidence indicates, as one would expect from an osteoimmune viewpoint, that bone formation and resorption are under the subtle control of multiple regulatory systems that include prostaglandins and leukotrienes (Pilbeam et al. 2002). Leukotrienes

are fatty molecules, naturally produced eicosanoid lipid mediators of the immune system that contribute to inflammation, and whose production generally accompanies the production of histamine for triggering and exacerbationg allergic reactions. In the presence of factors stimulating bone resorption, the production of PGE2 by COX-2 is favored in osteoblasts. Osteoclastogenesis is inhibited by the reduction of IL-1-induced COX2 activity and PGE2 production, in an essentially RANKL-independent process (Ha et al. 2006; Hiraga et al. 2006; Shoji et al. 2006). Additionally, LTB4 also favors osteoclastogenesis in a RANKL-independent manner (Traianedes et al. 1998; Anderson et al. 2001; Jiang et al. 2005).

In summary, therefore, it is becoming increasingly apparent that, as much as the emergence of immune cells via hematopoiesis occurs in bone, and through bone metabolism, products of the immune system influence bone metabolism, generation, and loss through resorption. It is also evident now that bone pathologies arise, more often than not, from immune reactions (e.g., inflammation), or engender immunopathologic responses, which in turn can exacerbate bone disease. The question that has become intertwined in this discussion at this juncture is, therefore, the manner in which the immune system might "talk" to bone metabolism.

1.2.2 The Immune System Talks to the Bone

Osteoimmunology, we now must all agree, represents a conceptual rethinking of multiple phenomena, interrelating biological events in bone and the immune system. The root of exploration of this interplay begins with the basic understanding that the bone environment is critical for the development of hematopoietic stem cells, from which the cells of the immune system derive, and that various immunoregulatory cytokines influence the fate of bone cells.

There is general agreement that lymphocytes influence bone remodeling by exerting an impact on osteoclastogenesis. Thus, T cells are assumed to be responsible for bone loss which occurs as a consequence of a series of pathological conditions, for example systemic viral infections and chronic local bone and joint diseases, such as rheumatoid arthritis or inflammatory bowel disease. Concerning the type of impact that T cells exert on osteoclastogenesis results from in vitro and in vivo experiments differ to a high degree. The same is true for different lymphocyte subpopulations, i.e., data concerning the effect of CD4 and CD8 lymphocytes on osteoclastogenesis, are not consistent.

On the one hand, data from literature suggest an inhibitory effect of T cells. In one in vitro study, Vit D-stimulated osteoclast-like cell formation was enhanced after lymphocyte depletion. This was attributed to increased PGE2 production and consecutive upregulated RANKL and downregulated OPG expression. IFN-γ was found to be the modulatory factor, which is produced by activated anti-CD3 T cells, and which interferes with TRAF6, thus strongly inhibiting the RANKL-induced activation of NF-κB and JNK in vitro. It was acknowledged that resting T cells

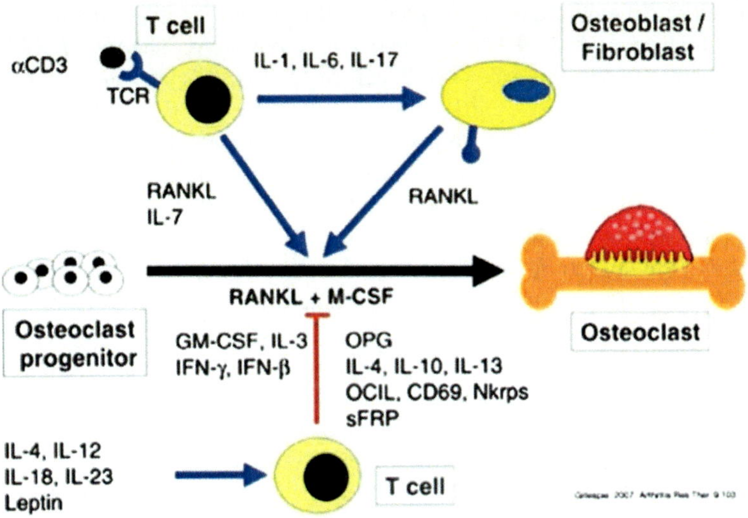

Fig. 1.3 Cytokines regulating T cells, osteoblasts, and osteoclasts. This schematic representation, adapted from Gillespie, 2007, shows to some degree the intimacy between the cellular immune system and bone metabolism. The central role played by RANKL in osteoimmune interaction is evinced

typically exert no effect on osteoclastogenesis (Grcević et al. 2000). T cells indeed have no effect in coculture with IFN-γR–/– BMMs (bone marrow-derived monocyte/macrophage precursor cells) stimulated by RANKL (Takayanagi et al. 2000).

By contrast to the above-mentioned results, there are cases where resting T cells may also negatively regulate osteoclastogenesis via production of granulocyte/monocyte colony-stimulating factor (GM-CSF) and IFN-γ by CD4 but not CD8 T cells (Shinoda et al. 2003). Another in vitro study demonstrated that the downregulatory effect of lymphocytes is due to the CD8 T cell subset, and effect at all independent of IL-4 and TGF-β (John et al. 1996) (Fig. 1.3).

Nonetheless, activated T cells promote and induce osteoclastogenesis both in vitro and in vivo. CD4 T cells stimulated by conjugated anti-CD3 with anti-CD28 costimulus exert their effect via membrane-bound and secretory RANKL, and inhibited osteoclastogenesis, whereas T cells activated with staphylococcal enterotoxin A, PHA, and Con A had inconsistent effects. The osteoclastogenic effect was CD4+ Tcell –dependent (Wyzga et al. 2004).

So the problem remains as to what degree can all the lines of evidence be taken as concerted and confluent, rather than contrasting and contradictory. Much fundamental research remains to fully characterize the cellular immune regulation of osteoclastogenesis.

- *Osteoblasts/stromal cells* are the main regulators of osteoclastogenesis. They express the cytokines RANKL, which binds to RANK on osteoclast precursors and thereby induces osteoclastogenesis, and OPG, which is able to prevent that interaction. Among others, TGF-β and 17-β-estradiol stimulate the production of OPG whereas Vit D, PTH, and PGE2 promote the production of RANKL. Upon activation via dendritic cells T cells activate osteoclasts directly through the secretion of sRANKL. Furthermore, T cells secrete INF-γ, which on the one hand stimulates macrophages to produce pro-inflammatory cytokines, which in turn promote RANKL expression in osteoblasts/stromal cells, and on the other hand suppresses permanent osteoclast activation by the destruction of TRAF6. But, these might be circumstantial and secondary to allied event, since, for example, we also know that endothelial cells have been shown to express RANKL and OPG and might therefore also participate, or perhaps even initiate and direct in the regulation of osteoclastogenesis (Yasuda et al. 1998).

- In the murine system, *T cells* do not seem to be absolutely required for osteoclastogenesis in a rheumatoid arthritis animal model, although they contribute to form an important pathologic feature in arthritic joints (Plows et al. 1999). These observations suggest either that the experimental animal model is not true in mirroring the human pathology, or that there exist fundamental species differences, perhaps deriving from the nonabsolute congruency between murine and human cellular immune processes, mechanisms, and cell populations and subpopulations. That said, it is unquestionable that cells in rheumatic joints are in a distressed state, attributable in part at least to IFN-γ. It follows therefore that pro-inflammatory cytokines act detrimentally upon synovial fibroblasts, across species and pathological model. These cells in the activated state – activated T cells, challenged synovial fibroblasts – are the main source of RANKL, which in turn is responsible for osteoclast differentiation, although RANKL has also been shown to be produced by T cells.
- *Cytokines* have a myriad of effects upon osteoclastogenesis: IL-1α, IL-1β, IL-6 and other members of the gp130 cytokine family, IL-7 (*vide infra*) and TNF-α directly or indirectly promote osteoclastogenesis, whereas IFN-β, IFN-γ, IL-3, IL-4, IL-10, IL-13, and IL-12 alone and in synergy with IL-18 inhibit osteoclast formation. The IFN-γ-mediated suppression of osteoclastogenesis, most likely occurs, as noted above, by inhibiting RANKL signaling by downregulation of the transcription factor TRAF6 expression via the signal transducers and activators of transcription family member 1 (Stat1). TGF-β, depending on the micro-environment, can both induce, via suppressor of cytokine signaling 3 (SOCS3) or suppress osteoclastogenesis (Theoleyre et al. 2004; Takayanagi et al. 2005). Osteoblasts may also serve in these activation processes since they posess antigen-presenting properties, express both MHC Class-II molecules and CD54 (ICAM-1) and CD166 (ALCAM), and are thus capable also of activating T cells. Osteoblasts express members of the TOLL-like receptor family (i.e., TLR-4, TLR-5, and TLR-9), indicating an active role in host immune response. Pattern recognition receptors were found not only on the surface of osteoblasts but also intracellularly (Marriott et al. 2005). The data could demonstrate the expression of the nucleotide-binding oligomerization domain proteins NOD1 and NOD2 following bacterial challenge of the cells. Osteoblasts also produce IL-6 upon encountering T cells and following stimulation by IL-17 (*vide infra*). Lastly, and of great relevance to osteopathologies, such as rheumatoid

arthritis, which often have an etiology that can be traced to a superantigen origin, osteoblasts are quite capable and endowed to present superantigen effectively to T cells. (Stanley et al. 2006). Nonetheless, there is relatively little detailed information on the cytokine production pattern of osteoblasts. IL-6 was shown to be produced by stromal cells/osteoblasts (Bordin et al. 2003). Production of IL-6 and RANKL by osteoblasts is promoted by PTH and TNF-α, but with markedly different kinetics. Whereas PTH induces a rapid, but transient elevation of both cytokines, TNF-α leads to a biphasic and sustained increase of these cytokines, thus indicating the potent role of TNF-α in osteoimmune pathologic conditions (Dai et al. 2006).

In brief, bone homeostasis refers to the constant process of remodeling, by which old bone is replaced by new bone. The skeleton, including the rostral skeleton is a metabolically active organ that undergoes continuous remodeling throughout development and aging. Two populations of cells drive this process: the bone-resorpting osteoclasts, and the bone-generating osteoblasts. Several factors control and regulate the process of bone homeostasis, including:

- Local factors, such as immune cell-produced cytokines (e.g., TGF-β; TNF-α; IL-1β; IL-12).
- Growth factors (e.g., M-CSF; BMP's).
- Mediators of cell-to-cell and matrix-to-cell communication.
- Products of the neuroendocrine system (e.g., estrogen, PTH; PTHrP; insulin-like growth factors [IGF's]) (Raisz 1999; Hadjidakis and Androulakis 2006).

As was discussed earlier and now can begin to fit into a *Gestalt*[3] of our understanding of the fiundamentals of oesteo-immunology. In brief, the RANK/RANKL/OPG system is critical to the processes of bone resorption and formation, are tightly coupled to allow the wave of bone formation to follow each cycle of bone resorption, and are the central engine that regulates bone and skeletal integrity (Hadjidakis and Androulakis 2006).

1. The regulation of bone metabolism occurs principally through the RANK signaling pathway. Stromal osteoblastic precursors express on their surface the RANK ligand (RANKL) as we have discussed earlier, and produce a soluble of RANKL, osteoprotegerin (OPG). Bone-resorbing hormones, including PTH, and cytokines, such as IL-1β or TNF-α, induce RANKL expression by stromal osteoblastic cells. RANKL then engages RANK on the osteoclastic progenitors, which induces their differentiation into osteoclasts. The process is enhanced by M-CSF, which is also produced by the osteoblast cells, and finds receptor-mediated binding and cell signaling on the maturing osteoclastic cells. OPG is a decoy ligand, preventing RANKL interaction with RANK on osteoclastic progenitors, and acts as a potent inhibitor of ostaclastogenesis.
2. The osteoimmune relationship notably manifests expression of RANKL by activated T cells, and T cells support osteoclastogenesis. OPG secretion is upregulated

[3] German for "shape," or totality of a given entity in terms of its shape. Used commonly to describe wholeness.

by anti-CD3 antibody stimulation of normal CD4 T cells in vitro, and enhanced by IL-4, IL-1β, TNF-α, GM-CSF, but blunted by IL-10 (Kotake et al. 2001; Colucci et al. 2004.

3. Other immune factors can substitute RANK/RANKL/OPG system in osteoclastogenesis. Bacterial lipopolysaccharide (LPS), the CD34 ligand, leads to inflammation systemically via IL-1β, IL-6, and TNF-α, and can stimulate osteoclastogenesis and bone resorption locally by binding to the Toll-like receptor 4, leading to NF-κB activation in OC. This effect is not blocked by OPG. Further, in combination with TNF-α, the inflammatory T-cell-derived cytokine found in fluid from osteoarthritic joints, IL-17 can stimulate OC differentiation and bone degradation in an NF-κB-independent process that also cannot be blocked by OPG.
4. The observation that T cells from arthritic joints also express RANKL suggests that IL-17 may be a significant pathway of osteoclastogenesis that is redundant with, but independent from the RANK/RANKL/OPG system (Fig. 1.4 and Table 1.1).

   - The *IL-17 family of cytokines* includes IL-17B, IL-17C, IL-17D, IL-17E (also called IL-25), and IL-17F. The primary function of IL-17-related cytokines is to modulate induction of many immune signaling molecules. The most notable

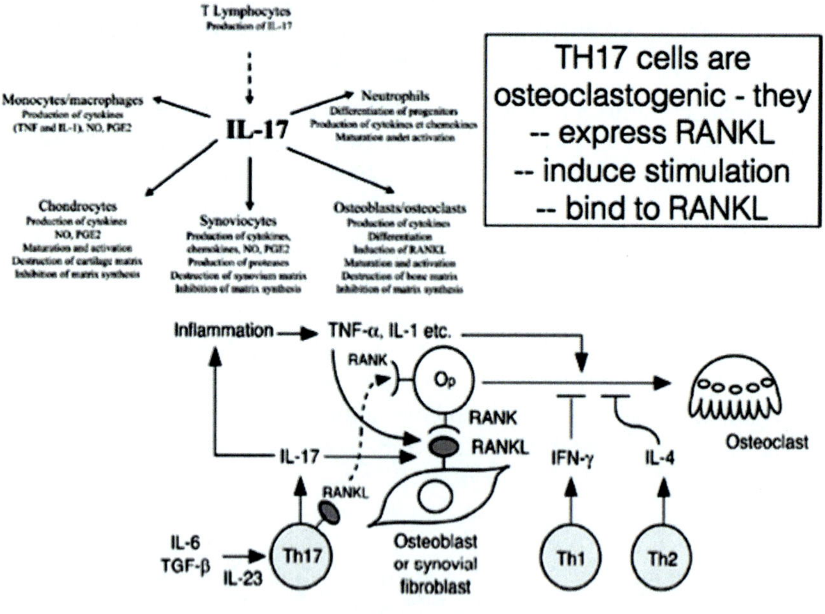

Fig. 1.4 The role of interleukin (IL)-17 in the osteoimmune system. This schematic representation of the multifaceted physiologic role of IL-17 shows that this inflammatory cytokine, produced by activated lymphoid-derived T cells rather than myeloid-derived monocytes/macrophages, contributes to the activation of the latter, as it also contributes to regulating the maturation of osteoclast precursors (Op) and, when further aided by T cell-produced TH1 and TH2 cytokines, the activation of bone resorbing osteoclasts. Much of the regulatory role of IL-17 is attributed to its modulation of RANKL. Adapted from: osteoimmunology.com/

Table 1.1 Modulators of RANKL, OPG, and RANK expression (taken from Weitzmanm and Pacifici 2007)

	RANKL	OPG	RANK
Hormones			
Vitamin D3[a]	↑	↑	↑
PTH	↑	↓	
PTHrP	↑		
Estradiol		↑	
Testosterone		↑	
Prolactin	↑	↓	
Cytokines			
TNFα	↑	↑	
TNFβ		↑	
IL-1α		↑	
IL-1β		↑	
IL-6[b]	↑	↑	↓
IL-11	↑	↑	
IL-17	↑		
CD40L	↑	↑	
Growth factors			
TGF-β	↓	↑	↓
BMP-2	↑	↑	
LIF	↑	↑	–
IGF-I	↑	↓	
VEGF			↑

role of IL-17 is its involvement in inducing and mediating proinflammatory responses. IL-17 is commonly associated with allergic responses. IL-17 induces the production of many other cytokines (such as IL-6, G-CSF, GM-CSF, IL-1β, TGF-β, TNF-α), chemokines (including IL-8, growth related gene alpha, GRO-α, and MCP-1), and prostaglandins (e.g., PGE2) from many cell types (fibroblasts, endothelial cells, epithelial cells, keratinocytes, and macrophages). The release of cytokines causes many functions, such as airway remodeling, a characteristic of IL-17 responses. The increased expression of chemokines attracts other cells including neutrophils but not eosinophils. IL-17 function is also essential to a subset of CD4+ T-Cells called T helper 17 (TH17) cells. As a result of these roles, the IL-17 family has been linked to many immune/autoimmune related diseases including rheumatoid arthritis, asthma, lupus, allograft rejection, and antitumor immunity. Each member of the IL-17 family has a distinct pattern of cellular expression. The expression of IL-17A and IL-17F appear to be restricted to a small group of activated T cells, and upregulated during inflammation. IL-17B is expressed in several peripheral tissues and immune tissues. IL-17C is also highly upregulated in inflammatory conditions, although in resting conditions is low in abundance. IL-17D is highly expressed in the nervous system and in skeletal muscle and

IL-17E is found at low levels in various peripheral tissues (Aggarwal and Gurney 2002; Kolls and Linden 2004; Yu and Gaffen 2008). Although it has only limited homology to other cytokines, IL-17 exhibits proinflammatory properties similar to those of TNFα, particularly with respect to induction of other inflammatory effectors, including several bone pathologies, most notably rheumatoid arthritis (Gaffen 2004). Research has established that signal transduction pathways dependent on PI3K/Akt and NF-κB are involved in bone-related pathology mediated IL-17 (Kim et al. 2005), possibly by modulating, in part at least, production of related pro-inflammatory cytokines (i.e., IL-6, IL-8) by synovial fibroblasts (Hwang et al. 2004).

- The *IL-17 receptor family* consists of five, broadly distributed receptors that present with individual ligand specificities. Within this family of receptors, IL-17R is the best described. IL-17R binds both IL-17A and IL-17F and is expressed in multiple tissues: vascular endothelial cells, peripheral T cells, B cell lineages, fibroblast, lung, myelomonocytic cells, and marrow stromal cells. IL-17RB, binds both IL-17B and IL-17E. Furthermore, it is expressed in the kidney, pancreas, liver, brain, and intestine. IL-17RC is expressed by the prostate, cartilage, kidney, liver, heart, and muscle tissues. The IL-17RC gene may undergo alternate splicing to produce a soluble receptor in addition to its cell membrane-bound form. In similar manner, the gene for IL-17RD may undergo alternative splicing to yield a soluble receptor. This feature may allow these receptors to inhibit the stimulatory effects of their yet-undefined ligands. The least-described of these receptors, IL-17RE, is known to be expressed in the pancreas, brain, prostate, and bone (Aggarwal and Gurney 2002; Gaffen 2004; Kolls and Linden 2004; Yu and Gaffen 2008).
- Lymphocytes expressing γδ T-cell receptors constitute an entire system of functionally specialized subsets that have been implicated in the regulation of immune responses, including responses to pathogens and allergens, and in tissue repair. γδ T cells represent a small subpopulation of T cells that, unlike aβ T cells, function more as cells of the innate immune system. γδ T cells are known to mediate the production of inflammatory cytokines, including interferon-γ, tumor necrosis factor-α, and interleukin (IL)-17, and thus enable the activation of other subsets of infiltrating effector cells. However, not much attention was paid to γδ T cells until the recent discovery of a distinct CD4+ T helper (TH) cell, TH17 cell. CD4+ T cells, upon activation and expansion, develop into different TH-cell subsets with different cytokine profiles and distinct effector functions. T cells were earlier divided into TH1 or TH2 cells, depending on the cytokines they produce. A third subset of IL-17-producing effector TH cells, called TH17 cells, has been discovered and characterized recently. Since then the literature on IL-17-producing cells has grown steadily, and several studies have focused on γδ T cells. Cytokine-mediated modulation of CNS inflammatory diseases by γδ T cells in humans or in animal models is currently the subject of many studies. IL-17 and its receptor IL-17R have been implicated in the pathogenesis of immune-mediated CNS diseases, and attention has been paid to understand the

mechanisms by which IL-17 cytokines and it receptor (IL-17R) family mediate the effects at a molecular level. This article reviews the studies that cover earlier aspects of γδ T cell/IL-17 biology and the new dimension of γδ T cells, IL-17, and IL-17/IL-17R signaling axis in CNS inflammation. Understanding the role of γδ T cells, IL-17, and IL-17/IL-17R signaling axis in infection and immunity could open a new avenue for immunomodulation (Das Sarma 2010; Barkhordarian et al. 2011).

- Immune cells that are endowed with the ability to respond to challenges by means of IL-17 are referred to as belonging to the TH17 group. TH17 cells are a subset of CD4+ T cells that are responsible for inflammatory and autoimmune disorders. Data demonstrate the presence of TH17 cells, some of which produce both IL-17 and IFNγ, indicating a putatively overlapping subpopulation of TH17/TH1 cells (Annunziato et al. 2007; Romagnani et al. 2009). In brief, TH17 cells are characterized by:
  - Surface expression of CCR6, IL-23R, IL-12Rb2, and CD161.
  - Expression of T-bet, retinoic acid-related orphan receptor (ROR)γτ.
  - Ability to produce IFN-γ and IL-17A in the presence of IL-12.
  - Ability to arise from CD161+CD4+ precursors, which constitutively express RORγτ and IL-23R, in response to the combined activity of IL-1β and IL-23.
  - Unresponsiveness to TGF-β for mediation of differentiation, although it can favor their proliferation by inhibiting T-bet expression (Romagnani et al. 2009).

1.2.3 The Osteoblastic Niche for Immunogenesis

The RANKL–RANK–OPG (osteoprotegerin) axis noted above is an example of an important signaling system functioning both in bone and immune cell communication. RANKL is expressed on osteoblasts and activated T cells, whereas RANK is expressed on osteoclasts, and dendritic cells, both of which can be derived from myeloid progenitor cells. Surface RANKL on osteoblasts as well as secreted RANKL provide necessary signals for osteoclast precursors to differentiate into osteoclasts. RANKL expression on activated T cells leads to dendritic cell activation through binding to RANK expressed on dendritic cells. OPG, produced by dendritic cells, is a soluble decoy receptor for RANKL that competitively inhibits RANKL binding to RANK.

The bone marrow cavity is important for the proper development of the immune system, and houses important stem cells for the maintenance of the immune system. Within this space, as well as outside of it, cytokines produced by immune cells also have important effects on regulating bone homeostasis. Some important cytokines that are produced by the immune system, including RANKL, M-CSF, TNFα, ILs, and IFNs, affect the differentiation and activity of osteoclasts and bone resorption. During chronic inflammation, the balance of bone modeling and remodeling can be greatly affected, contributing to painful and/or visible disorders in bone metabolism.

It is also critical to recognize at this juncture that the osteoblasts provide key factors for the development of hematopoietic stem cell niches. However, there is growing evidence that bone continues to play a role in adaptive immunity at later stages beyond lymphocyte development. For example, it is now known that long-lived memory T and B cells return to specialized niches in the bone marrow. The significance of this observation is currently unknown but could be important in the crosstalk between the bone and immune system.

Hematopoietic stem cells are located in the bone marrow and are responsible for the continuous production of blood cells in an adult organism. Their capacity for self-renewal and their ability to differentiate into multiple cell types is strongly dependent on their surrounding microenvironment, which is also referred to as stem cell niche. There, cells produce various signaling molecules, cell adhesion molecules, and components of the extracellular matrix and thereby determine the long-term repopulating ability of stem cells. Taichman and Emerson (1998) remarked that osteoblasts play a crucial role in stem cell maintenance due to an intimate cell-to-cell contact via integrins (Taichman et al. 2000). Another interesting observation was made in CBFA1 deficient mice, which are devoid of osteoblasts, and characterized by the absence of bone marrow. These mice do show normal hematopoietic development in ec topic sites, such as liver and spleen until day E17.5, suggesting an important role for osteoblasts in HSC homing into the bone marrow cavity (Ducy et al. 1997; Komori et al. 1997; Otto et al. 1997).

Bone homeostasis is in turn regulated by immune responses, particularly when the immune system has been activated by infection or becomes disregulated. In conditions such as periodontitis, infiltrating lymphocytes and other mononuclear cells produce key factors, which influence bone turnover by altering the balance between bone formation, mediated by osteoblasts, and bone resorption, mediated by osteoclasts. Beyond such pathological conditions, the question of whether the immune system influences normal oral bone metabolism, either by direct or indirect mechanisms, remains unanswered. However, the discovery of RANKL, and its characterization as a key differentiation factor for osteoclasts, and the findings that RANKL is expressed on activated T and B cells, has provided critical evidence for a potential link between normal immune responses and bone turnover. It is therefore becoming clear that crosstalk between the immune system and rostral bone through activated lymphocytes and bone cells occurs throughout life, as all mammals are constantly challenged by a diverse oral microflora, which induces some level of constant low grade immune system activation. Furthermore, as the aging process unfolds, there is an accumulation of memory T cells and B cells in the bone marrow, which express RANKL on their surface. These cells have now been shown to modulate bone turnover, in particular in periodontitis and other bone degenerating diseases.

Taken together, these and related issues place oral osteoimmunology in a position of unique clinical relevance.

1.3 Implication for Stomatology

1.3.1 Facial Osteology

The rostral or facial skeleton is complex and unique in terms of the nature and structure of the bones involved, their ossification during ontogenesis, and their articulation. It corresponds to the bones of the anterior and lower human skull, as opposed to the posterior skull, the neurocranium, which contains the brain and the brain stem. The facial skeleton contains the organs proper of the anterior aspect of the cranium (e.g., eyes, ears, nose, mouth, including tongue, tonsils, etc.), and is thus often referred to as the splanchnocranium or viscerocranium.

Eight bones form the neurocranium, which are juxtaposed by means of sutures that form immovable (i.e., synarthrodial) joints. Some degree of malleability and flexibility of the sutures is permitted by the Sharpey's fibers, named after the Scottish anatomist William Sharpey (1802–1880). These fibers are not unique to neurocranial sutures, and in fact present, for example, in the attachment of the periodontal ligaments; there are micro-anatomical histological structures that resemble matrices of connective tissue intertwined together to form bundles of strong collagenous fibers, which connect the periosteum to bone.

Fourteen bones form the splanchnocranium. Moreover, encased within the temporal bones are the six auditory ossicles of the middle ear. In addition, the hyoid bone, while supporting the larynx, does not articulate with any of the other cranial bones, and thus may, or may not be considered a component of the facial skeleton proper (Fig. 1.5).

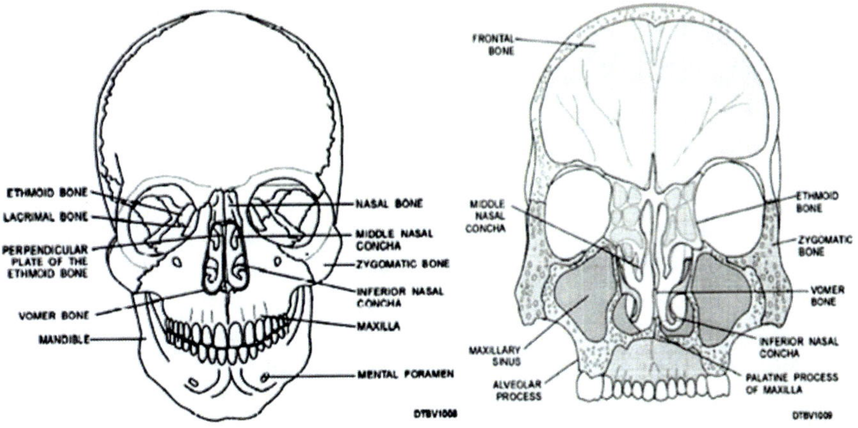

Fig. 1.5 The human facial skeleton. The figure shows a front view and a sagittal section of the facial skeleton from a human adult. Indicated are the principal bone structures that support the stoma superiorly (e.g., zygomatic, maxillary, vomer, palatine), and inferiorly (i.e., mandibular). The latter section present aspects of the deep face. Adapted from: http://www.tpub.com/content/medical/14274/

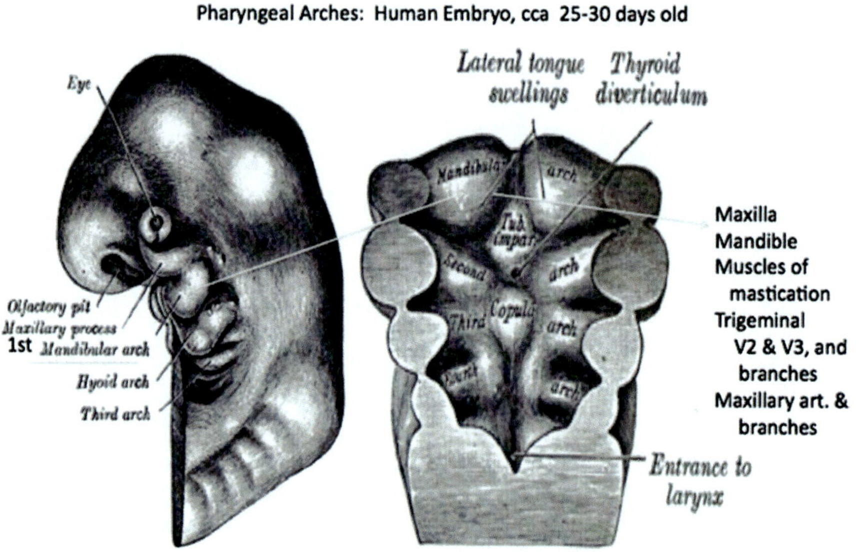

Fig. 1.6 Pharyngeal arches in the developing human fetus. The figure, adapted from Gray's anatomy, shows a side view and sagittal section of the anterio-frontal aspect of a human fetus at 25–30 days, gestational age. The specimen evinces clearly the first four tissue foldings located anterio-laterally to the median line and inferior to the developing stoma (i.e., olfactory pit). The first (mandibular) arch and associated lateral lingual tumerences are shown. The principal bones of the facial skeleton emerge from this arch. Source: Gray's Anatomy

During early ontogenesis, the bones of the facial skeleton derived from the pharyngeal arches (i.e., left and right mesodermal emergences, with no alterations in the ectodermal and endoderma layers, along the sides of the developing pharynx during the 3–5th week of intra uterine development in humans). There are six pharyngeal arches, but in human development, the fifth arch gives rise to no stomatological structures – hence, development proceeds along pharyngeal arches 1, 2, 3, 4, 6 as follows (Fig. 1.6):

- *First pharyngeal arch* – the mandibular arch: divides into a maxillary process to give rise to structures including the bones of the lower two thirds of the face (i.e., maxilla) up to and including the incus and malleus of the middle ear, as a superior regression of the Meckel's cartilage; and a mandibular process that comes to form a cartilaginous "model" of the mandible as a cartilaginous bar of the mandibular arch known as cartilage of Meckel (named after the German anatomist Johann Friedrich Meckel (1781–1833)), with the associated muscles of mastication, anterior belly of the digastric, mylohyoid, tensor tympani, tensor veli palatini, as well as the V2 and V3 branches of the trigeminal nerve and the corresponding branches of the maxillary artery. The mandible (i.e., "lower jaw") will eventually form by intramembranous ossification of the Meckel's cartilage.

- *Second pharyngeal arch* – the hyoid arch: development of the stapes ossicle of the middle ear, styloid process, hyoid (lesser horn and upper part of body), Reichert's cartilage, with the associated musculature of facial expression, buccinator, platysma, stapedius, stylohyoid, posterior belly of the digastric, as well as the facial nerve, and the stapedial artery.
- *Third pharyngeal arch* – this determines the development of the greater horn and lower part of body of the hyoid bone, as well as the stylopharyngeus muscle, the glossopharyngeal nerve, and the common and internal carotid arteries.
- *Fourth pharyngeal arch* – for the development of the thyroid and epiglottic cartilages, and the associated cricothyroid muscle, and all the intrinsic muscles of soft palate including levator veli palatini, the vagus nerve and the superior laryngeal branch, as well as the fourth aortic arches, on the right signifying the subclavian artery, and on the left signifying the aortic arch.
- *Sixth pharyngeal arch* – for the development of the cricoid, arytenoid, and corniculate cartilages, and associated intrinsic muscles of larynx except the cricothyroid muscle, the vagus nerve and its recurrent laryngeal branch, and the sixth aortic arches, on the right signifying the pulmonary artery, and on the left signifying the pulmonary artery and ductus arteriosus.

Clearly, with respect to the development of the oral cavity proper, the first pharyngeal arch, which is the first to form, separates the emerging mouth pit (i.e., stomodeum) from the pericardium, and forms pharingo-laryngeal structures. The stomodeum first arises as a depression between the developing structures that becomes the brain and the pericardium.

As the anterior aspect of the embryo forms and develops in the initial 1–4 weeks of ontogenesis, precursor rudiments of what soon develops into the cephalic flexure, the pericardial, and the bucco-pharyngeal membrane structures come to present on the antero- ventral surface of the embryo. As the brain further expands, and the forward bulging of the pericardium grows, the bucco-pharyngeal membrane forms a depression between these two prominences. This depression constitutes the stomodeum, which is lined by ectoderm, and is separated from the anterior end of the foregut by the membrane itself. The bucco-pharyngeal membrane is thus formed by the apposition of the stomodeal ectoderm and foregut endoderm. By the fourth week of intrauterine development, the bucco-pharyngeal membrane disappears, and a patent communication is established between the mouth and the emerging pharynx. The lips, teeth, and gums emerge from the ectodermal walls of the stomodeum. The tongue advances anteriorly from the floor of the pharynx: the anterior two thirds of the adult tongue derive from the first pharyngeal arch, and the posterior one third emerges from the hypobranchial eminence as the copula, formed by the forward growth and fusion of the ventral ends of the second, third, and part of the fourth arches.

Perinatally – whereas the principal features of the facial skeleton (cf., Fig. 1.5) are apparent, ontogenesis of the splanchnocranium is not finished, as stomatological ossification is not complete.

- *Inferior nasal concha*, as noted above (*vide supra*) extends laterally along the wall of the nasal cavity, and posteriorly to articulate with the conchal crest of the palatine bone. Anteriorly, it articulates with the conchal crest of the maxilla. The sphenopalatine (aka, pterygopalatine) ganglion, one of the four parasympathetic ganglion found in the deep face, lies in the pterygopalatine fossa that sits superiorly within it, and projects its branches to each of the superior, middle, and inferior ridges. The ganglion is suspended by nerve roots from the maxillary nerve, the second or middle branch of the trigeminal nerve (V2). The ganglion also has a sympathetic component, as it receives sympathetic efferent postganglionic fibers from the superior cervical ganglion. These fibers traveled from the superior cervical to the sphenopalatine ganglion via the deep petrosal nerve, which joins with the greater petrosal nerve, a branch of the facial nerve, to form the nerve of the pterygoid canal (aka, Vidian nerve, named after Vidus Vidius, the Italian anatomist born Guido Guidi, 1508–1569), that serves this ganglion. The ossification process of the nasal concha begins during the fifth month of intrauterine life from a single ossification center that emerges in the lateral wall of the cartilaginous nasal capsule.
- *Maxilla* is actually the product a fusion of two bones along the palatal fissure. On each side, the maxilla consists of the body of the maxilla, wherein lies the large maxillary sinus, the zygomatic, frontal, alveolar, and palatine processes, and the infraorbital foramen to permit passage of the infraorbital branch of the maxillary nerve (V2), and its associated artery. The maxilla articulates with two among the cranial bones (i.e., frontal and ethmoid), and several bones of the facial skeleton: nasal, zygomatic, lacrimal, inferior nasal concha, palatine, and vomer. It derives from the first pharyngeal arch, and its ossification is ossified from perhaps as many as six, or as few as two centers[4]: one for the maxilla proper and one for the premaxilla (i.e., *os incisivum, os intermaxillare*), which appear during the sixth week of intra uterine development. They eventually fuse completely (e.g., *sutura incisiva*), early in the third month. The suture line between the two maxillary portions may persist on the hard palate throughout adult life.
- *Palatine bone*, forms what is commonly known as the hard palate,[5] and continues posteriorly as the soft palate, which consists of a membranous aponeurosis and movable, fibromuscular fold, the velum palatini that is attached to the posterior edge of the hard palate, and extends postero-inferiorly to a curved free

---

[4] For an interesting account of the historical evolution of this area of research, cf., Barteczko and Jacob (2004).

[5] Note: the torus (pl., tori) palatinus (i) is a bony protrusion on the palate, most commonly found on the midline of the hard palate. Tori can also occur, abeit with a reduced prevalence, on the medial aspect of the mandibular bone (torus mandibularis). Palatal tori are more prevalent in the Asian, compared with the white Anglo-saxon US populations (20–35%). The condition is typically more prevalent in the young adult population, and tori may reduce in size in aging because of bone resorption events. The condition is twice more common in females, but relatively similar between blacks and whites in the United States. They may be an autosomal dominant trait.

margin from which hangs a conical process, the uvula. The palatine aponeurosis provides attachment for the four muscles of the soft palate: the Levator Veli Palatini, Tensor Veli Palatini, Palatoglossus, Palatopharyngeus, as well as the muscle of the uvula, the Musculus Uvulae. The palatine bone articulates with the bones of the deep face, important for the internal structure of the oral cavity, the sphenoid, ethmoid,[6] maxilla, inferior nasal concha, vomer bones. The palatine bone contributes to the floor and lateral wall of the nasal cavity, the roof of the mouth, and the floor of the orbit. Furthermore, it contributes to the formation of the pterygopalatine and pterygoid fossæ, and the inferior orbital fissure. The palatine bone is ossified in membrane from a single center, which makes its appearance about the sixth to eighth gestational week. The center appears at the angle of junction of the two parts of the bone, and spreads medially to the horizontal face, inferiorly into the pyramidal process, and superiorly to the vertical facet. These events lead to a putative timeline of ossification of the palatine bone as proceeding forward from four centers that are responsible for

- The pyramidal process and portion of the vertical part behind the pterygopalatine groove
- The remaining vertical and the horizontal portions
- The orbital process
- The sphenoidal process

At birth, the height of the vertical aspect of the palatine bone is roughly equal to its transverse width's horizontally part, and during normal postnatal ontogenic development, the former progressively grows larger than the latter.

- *Vomer bone* is a thin, somewhat quadrilateral, or trapezoidal bone situated in the median plane, and forms the inferior and posterior aspects hinder of the nasal septum. Along its two surfaces run the nasopalatine groove obliquely downward and forward to aid the nasopalatine nerve and vessels. The vomer articulates with the sphenoid and ethmoid bones, the two maxillae, and the two palatine bones, and specifically its inferior border articulates with the crest formed by the maxilla and palatine bones. The vomero-nasal organ, also called Jacobson's organ, is a chemoreceptor organ named for its closeness to the vomer and nasal bones, and is particularly developed in felines and canines. Early in ontogenesis (fifth to seventh week of gestation), the septum of the nose consists of a plate of cartilage, the ethmo-vomerine cartilage. As the postero-superior part of this cartilage is ossified, it comes to form the perpendicular plate of the ethmoid. Its antero-inferior portion persists as the septal cartilage. By contrast, the vomer is ossified in the membrane covering its postero-inferior part. Two ossific centers, one on either side of the middle line, appear about the eighth gestational week, which generates each of the two lamellae of the vomer, and which eventually joins and fuses about the third month of fetal life. A deep groove remains, which

[6] Both the ethmoid and the sphenoid bones are, as discussed above, superior and deep to the oral cavity proper, and serve more properly the nasal, rather than the oral stomata.

retains the cartilage, which is progressively absorbed in the continued process of ossification during ontogenesis as the union of the lamellae extends upward and forward. Postnatally, it takes another 10–15 years of growth and development before the lamellae is completely united to form the median plate of the vomer. The evidence of the bi-laminar origin of the vomer bone remains in adulthood in the everted alae of its upper border and the groove on its anterior margin.

- *Zygomatic bones* in vertebrates are small quadrangular paired bones which is present on each side of the face socket, forming the prominence of the cheek. They are also called in lay term, the cheekbones (aka, malar [Lt., malaris, cheeckbone] bones, malar-temporal bones). They play a critical structural role as they articulate with the maxilla, the temporal bone, the sphenoid bone, and the frontal bone. The malar aspect of the zygomatic bone presents the zygomaticofacial foramen for the passage of the zygomaticofacial nerve and vessels; the temporal aspect of the zygomatic bone supports articulation with the maxilla, and forms the anterior boundary of the temporal fossa, the lower a part of the infratemporal fossa, an irregularly shaped cavity, situated below and medial to the zygomatic arch, and bounded laterally by the ramus of mandible. During ontogenesis, the zygomatic bones ossify from three centers: the malar aspect, the superior and inferior aspects of the orbital facets, starting about seventh to eighth gestational week, and resulting in fusing by the fifth month of intrauterine life.
- *Mandible* articulates with the two temporal bones at the temporomandibular joints. These are complex synovial joints (i.e., gliding or arthrodial) whose motion, while considerably occurs in one plane only (i.e., hinge or gingymal), and which suffer from several dysfunctions, including inflammatory osteoarthritis, with significant local and systemic *sequelae*. The mandibular bone has a large and deep medullary core, where osteoimmune processes, such as those described earlier occur, as well as a cortical rim that is 2–4 mm thick is found. The ontogenic ossification of the mandibular bone is complex, and is initiated within the fibrous membrane covering the outer surfaces of the two Meckel's cartilages (right and left). The proximal (i.e., cranial) ends become contiguous with the ear capsules, as their distal extremities conjoin at the symphysis by mesodermal tissue. The process of ossification runs forward immediately inferior to the condyles and eventually incline superior-medially to the symphysis. Proximally, the malleus and incus, two of the bones of the middle ear, arise. The next succeeding portion of the cartilage, as far as the lingula, is progressively replaced by fibrous tissue which forms the sphenomandibular ligament. As the cartilage disappears between the lingula and the canine tooth; ossification also proceeds inferior-posteriorly to where the incisor teeth sits. In brief, ossification of the mandibular bone from the Meckel's cartilage is considered to arise independently on the right and on the left side from a single center, which appears near the mental foramen about the fifth to seventh week of gestation. Accessory nuclei of ossification from the cartilage may appear later in prenatal life to bring this complex and multifaceted process forward, but they possess no separate ossification centers. Rather, they are invaded by the surrounding membrane bone, which engenders the process of absorption. By contrast, the inner alveolar border of the

mandibular bone may arise from a separate and distinct ossification center during ontogenesis, the splenial[7] center, which results from an ingrowth from the main mass of the bone. Perinatally, the mandibular bone consists of two parts that are joined by a fibrous symphysis, which completes its process of ossification in the first year postnatally. The condylar process is positioned in the superior-lateral most aspect of the ramus of mandibular bone, which signifies the attachment of three of the four powerful masticatory muscles (i.e., masseter, temporal, medial pterygoid). The condylar and pericondilar aspects of the mandibular bone (e.g., subcondylar region located between the condyle and the coronoid process of the mandible; aka, mandibular notch) are most prone to fracture (36%) during post-natal development and adult life. These are dangerous fractures, which require delicate interventions, because they can engender significant and profound dysfunctions of the temporo-mandibular joint (*vedi infra*), consequential inflammatory events that may precipitate osteoarthritis of the jaw joint. It is also the case, however, and it is important to note that traumatic arthritis without condylar fracture may also develop from indirect transmitted violence to the superior aspect of the ramus of the mandibular bone. From an osteoimmune perspective, condylar fractures are critical. Condylar fractures can be

- Extracapsular
- Subcondylar
- Intracapsular

The powerful lateral pterygoid muscle then tends to cause important anterior and medial displacement of the condylar process (i.e., *processus condyloideus*), which signify the onset of joint disorders, including damage to branches of the trigeminal and facial cranial nerves that run proximally, either lateral or medial, to the condylar and pericondylar aspects. Several types of condylar fractures are consequential to trauma (e.g., accidents, sports), and present in order of increasing severity as follows:

- *Type I*: fracture of the neck of the condyle with relatively slight displacement of the head. The angle between head and axis of ramus can vary from 10 to 45°.
- *Type II*: fracture that produces an angle from 45 to 90°, and that results in tearing of the medial portion of the joint capsule, with potential associated hemarthrosis.
- *Type III*: fracture that is so severe that the fragments are not continuous anymore, but still confined within the area of the glenoidfossa, resulting in significant medial and anterior displacement, and a tearing of the capsule such that the condylar head is now outside the capsule, and associated hemarthrosis. Often, in these cases, the fracture is associated with partial or complete

[7] Derived from splenia (Lt. pl of *spenium*, splint). It is a piece found in the mandibular bone in *homo sapiens*, and well retained throughout phylogeny as it is evident in facial skeletons from dinosaurs, reptiles, birds, and early mammals as well. It serves as a splint typically along the ventromedial surface of the mandible.

rupture of inferior dental artery, which signifies impaired endosteal blood supply and endogenous osteoimmune healing processes.
- *Type IV*: fracture of the condylar head that now comes to articulate on or in a forward position with regard to the articular eminence.
- *Type V*: fracture that presents vertically or obliquely through the head of the condyle.

Hyoid bone rests at the level of the base of the mandible in the front and the third cervical vertebra behind, and provides attachment to the muscles of the floor of the mouth and the tongue above (i.e., Middle pharyngeal constrictor, Hyoglossus, Digastric, Stylohyoid, Geniohyoid, and Mylohyoid muscles), and the larynx below, and the epiglottis and pharynx behind (i.e., Thyrohyoid, Omohyoid, and Sternohyoid muscles). It is a suspended bone held in place by the thyroid ligaments. Thus, it does not articulate with any other, and functions to allow a wider range of tongue, pharyngeal, and laryngeal movements by bracing these structures alongside each other in order to produce variation. Anatomically, it consists of a central part, the body (i.e., *corpus oss. Hyoidei, aka basihyal*), that sits antero-medially, and two pairs of latero-posterior extensions, called the cornua. The greater cornu (i.e., *cornua majora; aka thyrohyals*) project backward from the lateral borders of the body, flatten from above downward, and diminish in size from anterior to backward. They project into a tubercle that attaches the lateral hypothyroid ligament. The lesser cornu (i.e., *cornua minora, aka ceratohyals*) are smaller conical eminences, that are attached by their bases to the angles of junction between the body and greater cornua by fibrous tissue, and occasionally distinct diarthrodial joints (i.e., synchondroses in early ontogeny, and after middle life usually by bony union), which persist throughout ontogeny, but are at risk of become ankylosed (*vedi infra*) in later life. The early ontogeny of the hyoid bone is complex in that the lesser cornu and the most superior aspect of the hyoid body originate from the second pharyngeal arch, while the greater cornu and the lower part of the body of hyoid arise from the third pharyngeal arch. Moreover, ossification of the hyoid bone proceeds from six centers: two for the body, and one for each cornu. The process of ossification commences in the greater cornua during late fetal life, and proceeds rapidly to the body, and terminates in the lesser cornua in the first to second year postnatally. The connection between the body and greater cornu may remain fibrous until middle-age.

1.3.2 The Temporomandibular Joint

The temporomandibular joint (i.e., *articulatio temporomandibularis*; aka, "jaw joint"; TMJ) forms the articulation of the upper temporal bone, which is part of the cranium, superiorly, and the mandibular bone, commonly called the mandible (i.e., the "lower jaw"), inferiorly. The TMJ is a synovial joint, which is nearly unique, with the sternoclavicular joint, in that it possesses an articular disc that cushions the joint, composed of fibrocartilagenous tissue (i.e., firm and flexible elastic cartilage). From an anatomical perspective, the disc creates two spaces:

- *Inferior* to the articular disc, a TMJ compartment is formed by the mandible and the articular disc, involved in a rotational movement, which is the initial movement of the jaw as the mouth opens. The rotation of the condylar head around its instantaneous axis of rotation permits the first 20 mm or so of the opening of the mouth. Beyond that point, a translational movement[8] is necessary for further extension of the joint, which is achieved by the superior compartment of the TMJ.
- *Superior* to the disc, the TMJ compartment consists of the articular disk and the temporal bone, and is involved in the translational movements of the jaw, which corresponds to the secondary gliding motion of the jaw as the mouth is opened widely. The superior compartment is divided into two parts by a narrow slit, the petrotympanic fissure (Glaserian[9] fissure, named after the Swiss anatomist, Johann Glaser (1629–1675)).

The part of the mandible that conjoins to the under-surface of the disc is the mandibular condyle (*vedi supra*). The aspect of the temporal bone that conjoins with the upper surface of the disk is the glenoid (or mandibular) fossa, a concave depression in the squamous portion of the temporal bone that is bounded anteriorly by the articular tubercle, and posteriorly by the tympanic part of the bone, beyond which lies the external acoustic meatus. That proximity to the auditory organ provides some degree of explanatory rationale for why, as osteoimunnological events precipitate TMJ dysfunctions (*vide infra;* aka temporomandibular dysfunctions, TMD), one common complaint is ringing in the ears (i.e., *tinnitus*[10]). Those relationships develop and establish in ontogenesis around 12 weeks *in utero*, ossification proceeds (*vide supra*), and the joint spaces and the articular disc develops.

[8] Translation of the TMJ has been traditionally seen as a forward and downward sliding motion, on the anterior concave surface of the glenoid fossa and the posterior convex surface of the articular eminence. However, it is now clear that this translation actually amounts to a rotation around another axis, and that it engenders an "evolute," the resultant axis of mandibular rotation, which lies in the vicinity of the mandibular foramen, allowing for a low-tension environment for the vasculature and innervation of the mandible (Moss 1972).

[9] Not to be mistaken with the term "Gasserian," such as in the gasserian (i.e., trigeminal, aka "semilunar") ganglion, a sensory ganglion of the trigeminal nerve the Meckel's cavity within the dura mater, and that covers the trigeminal impression near the apex of the petrous part of the temporal bone, whence emerge the ophthalmic, maxillary, and mandibular branches of the trigeminal nerve, Cranial nerve V. The gasserian ganglion was named by Anton Hirsh, student of Johann Lorenz Gasser (Austrian anatomist, 1723–1765), in Anton Balthasar Raymund Hirsch. *Pars quinti encephali disquisitio anatomica*. Vienna, 1765. Re-published by Christian Friedrich Ludwig (1751–1823): *Scriptores neurologici minores selecti, sive opera minora ad anatomiam, physiologiam et pathologiam nervorum spectantia*. Volume 1, pp. 244–162. Leipzig, 1791.

[10] To be sure, tinnitus, the perception of ringing or sound within the ear in the absence of corresponding external sound, may result from a range of underlying causes, ranging from ear infections, to foreign objects or wax in the ear, nose allergies that prevent (or induce) fluid drain and cause wax build-up, and side effect of medications, of genetic (congenital) hearing loss, or of noise-induced hearing loss.

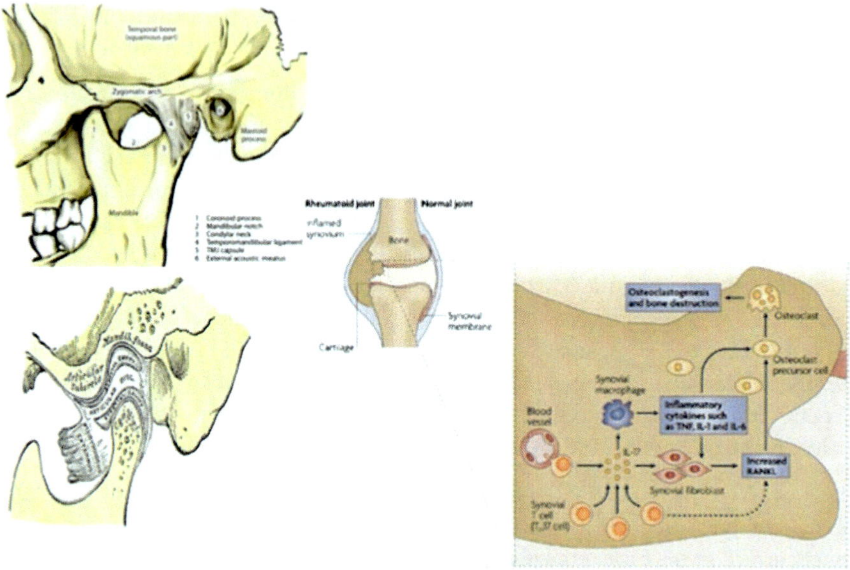

Fig. 1.7 Rheumatoid arthritis in the temporomandibular joint A side and a sagittal view of the temporomandibular-joint are shown, the latter presenting the position of the synovial articular disk between the mandibular condyle and the maxillary articular tubercule. The schematic representation outlines the events leading from inflammation (i.e., IL-17 regulated) to osteoclastogenesis and bone resorption. This process distinguishes the arthrtic joint, with inflammed synovium, from a normal joint characterized by a normal synovial membrane lining. Pathway by which RANKL activates osteoclast bone resorption and synoviocytes, which then invade cartilage is not shown. Takayanagi (2007, 2009); source: http://www.nature.com/nri/journal/v7/fig_tab/nri2062_F2.html

The TMJ is a two-component joint in structure, as outlined above, and function. It is supplied by branches of the external carotid artery, such as the superficial temporal branch, the deep auricular artery, the anterior tympanic artery, the ascending pharyngeal artery, and the maxillary artery. Functionally, the TMJ is a ginglymo-arthrodial joint, and both the articulation in one plane[11] and the gliding[12] properties of the joint can, and will suffer significantly in the case of osteoimmune pathologies, such as osteoarthritis, osteoporosis, or osteopetrosis (*vide infra;* cf. Fig. 1.7). Within these spatial constraints, the TMJ has several movements:

- *Excursions*: normal movements of the mandible during function, such as mastication, or chewing. Here, the working side condyle, which lies on the side

[11] A ginglymal joint is one in which the articular surfaces are molded to each other in such a manner as to permit motion only in one plane.

[12] An arthrodial joint is a synovial joint which, under physiological conditions, allows only gliding movement.

of the mandible that moves outward, performs rotation in the horizontal plane, whereas the balancing side condyle performs translation.

- *Protrusion*: a specific type of nonlateral, but rather forward excursion. Protrusion is accomplished by translation of the condyle down the articular eminence superiorly, without any more than the slightest amount of rotation taking place in the inferior space of the TMJ, other than that necessary to allow the mandibular incisors to come in front of the maxillary incisors without running into them.
- *Retrusion*: a reversal of protrusion.

The physical movements of the TMJ are directed by the four powerful muscles of mastication, and specifically the masseter, medial pterygoid, lateral pterygoid, and temporalis muscles. These muscles are innervated by the mandibular division of the trigeminal nerve (V3) and its sensory branches (i.e., auriculotemporal and masseteric branches of V3), and work to permit the mandible to move in different directions.

1. *Lateral pterygoid* muscle acts to pull the disc and condyle forward within the glenoid fossa, and down the articular eminence (i.e., open the jaw).
2. *Masseter* and *medial pterygoid* muscles close the jaw by pulling up the angle of the mandible.
3. *Temporalis* closes the jaw by pulling up on the coronoid process.

The stability of the TMJ is ensured by fibrous ligaments, which on the one part provide a substantial amount of flexibility and strength, but on the other often harbor inflammatory processes, which may precipitate sustain osteoimmune pathologies that can be significantly damaging to the TMJ and produce a serious impairment in the quality of life of the patient (*vide infra*). Ligaments that define the border movements of the TMJ include:

- The *temporomandibular ligament*, the thickened lateral portion of the capsule composed of an outer oblique and an inner horizontal portion (IHP).
- The *stylomandibular and sphenomandibular ligaments* are accessory and not directly attached to any part of the joint. The former separates the infratemporal region (anterior) from the parotid region (posterior), and runs from the styloid process to the angle of the mandible. The latter runs from the spine of the sphenoid bone to the lingula of mandible.
- *Oto-mandibular ligaments* that connect the middle ear (i.e., malleus) with temporomandibular joint (e.g., discomallear, and malleomandibular ligaments).

1.3.3 Immuno-Biology of Facial Bony Structures

As one considers the immuno-biology of facial bones, the question arises as to whether or not the osteologic structures that constitute the facial skeleton, and which we reviewed above, indeed contain adult stem cell niches, or are at all endowed with significantly relevant extent bone regeneration. Understandably, that question is clinically relevant for oral surgeons, implantologists, and oral biologists in general.

Research findings over the past few years converge to the widely accepted view that, in fact, bone marrow stem cells have the potential to recreate tissues of the craniofacial region to restore normal structure and function in reconstructing the hard tissues of a face (Robey and Bianco 2006). Indeed, postnatal skeletal stem cells are a subpopulation of the bone marrow stromal cell network, and in fact only 10%, at most, of the bone marrow clonal strains can form bone as definite adult skeletal stem cells (Bianco et al. 2006).

Whereas they are fundamentally similar to other bones, the osteological structures that constitute the splanchnocranium demonstrate discrete responses to developmental, mechanical, and homeostatic regulatory signals. Adult bone marrow stem cells obtained from either the mandible or long bones differentiated into osteoblasts, but mandible-derived cells exhibited a distinct epigenetic program, as manifested by increased expression of specific osteoblastic proteomic signature (e.g., alkaline phosphatase activity, mineralization, osteoblast gene expression), and overall greater effectiveness in the formation of colonies and larger bone nodules containing significantly more mineralized bone compared to osteoblasts derived from long-bone bone marrow stem cells. Taken together, the data to date suggest that mandible bone marrow-derived stem cells are endowed with increased osteogenic potential and augmented capacity to induce bone formation in vitro and in vivo, compared with the potential of bone marrow-derived osteoblasts obtained from any long bone of the articular skeleton (Aghaloo et al. 2010).

Case in point, in situations of bone atrophy and pneumatization of the maxillary sinus consequential to loss of teeth in the posterior maxilla, the dimension of alveolar ridge is decreased and sinus augmentation procedures are to create bone quantity and quality to ensure the placement of dental implants (Park 2010). Failure to provide adequate support at the posterior dentition will lead to a series of occlusion issues, including loss of vertical dimension of the temporomandibular joint (TMJ), with possible compression of the auriculo-temporal nerve and local trigeminalgia, as well as general trigeminalgia via the Gasserian ganglion (Demerjian et al. 2010). The fact that adult stem cells, derived from various tissues including bone marrow, periosteum, and trabecular bone, have been successful in sinus augmentation procedures both experimentally and clinically across the board (Park 2010), together with the observation that bone morphogenetic proteins (*vide infra*, BMPs) manifests osteoinductive properties in this context (Park 2009), attests to the osteobiological and osteoimmune resiliency of the facial skeleton.

As discussed earlier, BMPs and osteogenic proteins (OPs) are pleiotropic members of the transforming growth factor (TGF)-β supergene family, and as such play a central role in the osteoimmune network discussed here. These proteins act as soluble signals for the de novo initiation of bone formation, for sculpting the multicellular mineralized structures of the bone–bone marrow organ, and, in the context of the facial skeleton, not only favor the induction of cementogenesis, but also induce the morphogenesis of the periodontal ligament system with a faithful insertion of Sharpey's fibers into the newly formed cementum. In the facial bony structure, as in the skeleton in general, these OPs of the TGF-β superfamily contribute to sculpting tissue constructs that engineer skeletal tissue regeneration in molecular terms:

in brief, they regulate the induction of bone, such that, as bone develops a mosaic structure, cytokine members of the TGF-β superfamily singly, synergistically, and synchronously initiate and maintain tissue induction and morphogenesis (Ripamonti et al. 2005). Direct translational implications of these basic findings of fundamental oestoimmunology is the clinical observation that clefts of the anterior maxilla can undergo complete osseous regeneration induced by recombinant human bone morphogenetic protein type-2 (rhuBMP-2) (Herford et al. 2007).

Furthermore, distraction osteogenesis is a fundamental event of the cranio-maxillo-facial reconstruction process. The molecular mechanisms that regulate bone synthesis in the interfragmentary gap resulting from the gradual separation of bone segments, which signify the fundamental histology and cytology associated with distraction osteogenesis, are driven in large part by the RANK/RANKL/OPG system, discussed earlier (*vide supra*) for its central role in regulating bone metabolism and osteoclastogenesis. Taken together, the data to date strongly support an important biological and molecular role for the influence of the RANK/RANKL/OPG system on the bone healing and remodeling processes in distraction osteogenesis, and suggest possibly developing molecular interventions directed at improving the clinical outcome for distraction osteogenesis[13] (Pérez-Sayáns et al. 2010).

[13] Distraction osteogenesis, a surgical intervention introduced by the Italian surgeon Alessandro Codivilla (1861–1912) is used to reconstruct skeletal deformities primarily of the long bones, and is also applied to correct deformities of the maxillary and mandibular bones. When successful, the intervention simultaneously increases bone length and the volume of surrounding soft tissues. Maxillofacial surgeons use distraction osteogenesis for the correction of micrognathia, midface, and fronto-orbital hypoplasia in patients with craniofacial deformities.

Chapter 2
Osteoimmunopathology

2.1 Inflammatory Processes

2.1.1 Osteoarthritis

Osteoarthritis (cf., Fig. 1.7) also known as degenerative arthritis or degenerative joint disease is a group of mechanical abnormalities involving degradation of joints, including articular cartilage and subchondral bone. Symptoms may include joint pain, tenderness, stiffness, locking, muscle atrophy, and sometimes an effusion. Osteoarthritis may find its etiology in hereditary, developmental, metabolic, mechanical trauma, and progressive loss of cartilage.

Osteoarthritis is the most common form of arthritis. An estimated 15% of Americans suffer from arthritis, and the annual cost to society is estimated to be $95 billion. It is the second most common cause cited in claims for Social Security disability benefits.

Primary Osteoarthritis (i.e., nodal Osteoarthritis; erosive or inflammatory osteoarthritis) is a chronic degenerative disorder related to but not caused by aging. As biological age progresses, the water content of the cartilage decreases as a result of a reduced proteoglycan content. This outcome results in the cartilage to be less resilient, and more susceptible to degradation and degeneration. Inflammation of the surrounding joint capsule ensues as breakdown products from the cartilage are released into the synovial space. New bone outgrowths, bone spurs (aka, osteophytes), form on the margins of the joints, possibly in an attempt to improve the congruence of the articular cartilage surfaces. These bone changes, together with the inflammation, can be both painful and debilitating.

Secondary osteoarthritis can be caused by other factors but the resulting pathology is the same as for primary osteoarthritis:

- Congenital disorders of joints, including Marfan syndrome
- Diabetes and obesity
- Inflammatory diseases, for example

F. Chiappelli, *Osteoimmunopathology: Evidence-Based Perspectives from Molecular Biology to Systems Biology*, DOI 10.1007/978-1-4419-9495-0_2,

- Perthes' disease (i.e., degenerative disease of the hip joint, with idiopathic avascular osteonecrosis of the capital femoral epiphysis of the femoral head leading to an interruption of the blood supply of the head of the femur close to the hip joint)
- Lyme disease (Lyme arthritis is usually observed in the joints of the extremities: predominantly the knees, ankles, elbows, and wrist, and secondarily the hips and shoulders)
- Other chronic forms of arthritis (e.g., costochondritis [i.e., Tietze syndrome], gout, and rheumatoid arthritis)

• Injury and sepsis to joints, with possible ligament deterioration or instability may be a factor, etc.

The clinical success of tumor necrosis factor (TNF)-α blocking biologics in a growing number of immune-mediated pathologies, including, with respect to bone pathology, rheumatoid arthritis, and Ankylosing spondylitis. These and other non-osteoimmune conditions, such Crohn's disease and psoriasis, confirm the importance of TNF-α in driving chronic inflammation. This realization represents an important step forward in the treatment of these conditions (vide infra). TNF-α blockade, however, is a treatment, rather than a cure, and is not effective in all patients or in all autoimmune diseases and further research is needed to get closer to a cure. Recently, the identification of the novel IL-17 producing T helper cell subset as noted earlier (*vide supra*) that plays a dominant pathogenic role in animal models of autoimmunity, is a major advance on existing knowledge; although the role of these cells in human disease remains to be established. Cytokines driving angiogenesis are also important in disease chronicity and thus might be valid therapeutic targets (Williams et al. 2007).

IL-17-producing helper T cells of the TH17 subpopulation play a major role in the etiology and progression of this osteoimmune pathology by RANKL (receptor activator for nuclear factor ÐB ligand), and thus favoring osteoclastogenesis, and soted above. IL-12 and IL-18 participate in inflammatory processes that can lead to highly destructive osteolysis, yet these cytokines potently block osteoclast formation through mediation of T cells. IL-23 participates in chronic inflammatory processes, but lack of this cytokine results in reduced bone mass in mice, pointing to an influence on physiological regulation of bone mass.

2.1.2 Osteoporosis in HIV– and HIV+ Persons

2.1.2.1 Osteoporosis in Otherwise Healthy Adults

Osteoporosis develops when bone that is lost is not replaced by new bone. It results from any imbalance of bone turnover that results due to excess of osteoclast activity (bone resorption) over osteoblast activity (bone formation). Biochemical markers that reflect bone remodeling in osteoporosis and which can be measured in blood or urine include bone resorption markers (cf., Table 2.1), such as pyridinoline,

Table 2.1 Biomarkers of bone turnover (taken from Indumati and Patil 2010)

Markers of bone formation	Abbreviation	Sample used for analysis
Total alkaline phosphatase	Total ALP	S
Bone-specific alkaline phosphatase	Bone-ALP	S
Osteocalcin	OC	S
Undercarboxylated Osteocalcin	ucOC	S
Procolagen-I extension peptides		
N-Propeptide	PINP	S
C-Propeptide	PICP	S
Markers of bone resorption		
Hydroxyproline	Hyp	U
Hydroxylysine	Hyl	U
Galactosyl hydroxylysine	Gal-Hyl	U
Glucosyl galactosyl hydroxylysine	Glc-Gal-Hyl	U
Pyridinoline (total free peptide-bound)	PYD	U
Deoxypyridinoline (free, total)	DPD	U
Type I collagen telopeptides	NTX	S/U
N-telopeptide (N-terminal cross-linking telopeptide)	CTX	S/U
C-telopeptide (C-terminal cross-linking telopeptide)	CTX-MMP	U
C-telopeptide generated by matrix metallo proteinases	BSP	S
Bone sialoprotein	ACP	S
	TRACP	S
Acid phosphatase		
Tartarate resistant acid phosphatase		
(5b isoform osteoclasts)		

deoxypyridinoline, and collagen cross-links, as well as bone formation biomarker, among which the most notable (*vide supra*) are: alkaline phosphatase and osteocalcin (Indumati and Patil 2010).

Osteoporosis is a disease of bones that leads to an increased risk of fracture. It is among the most common human bone disease and is characterized by low bone mass or bone mineral density and loss of bone tissue such as: disrupted bone microarchitecture, and the altered amount and variety of proteins in bone. It is defined by the World Health Organization (WHO), in women, as bone mineral density of 2.5 standard deviations below peak bone mass (20-year-old healthy female average).

Osteoporosis is most common in women after menopause (i.e., postmenopausal osteoporosis), but may also develop in men or women of any age consequential to hormonal disorders (e.g., hyperparathyroidism) and other chronic diseases, or glucocorticoids intake (i.e., steroid- or glucocorticoid-induced osteoporosis). Given its influence in the risk of fragility fracture, osteoporosis may significantly affect life expectancy and quality of life. Osteoporosis can be prevented with lifestyle changes (e.g., exercise) and sometimes medication (e.g., dietary supplements of calcium, VitD, bisphosphonates) (Fig. 2.1).

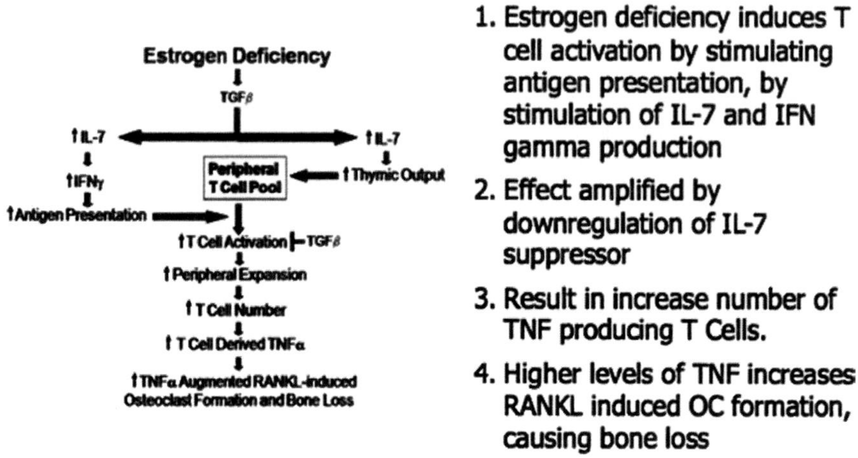

Fig. 2.1 Estrogen deficiency and bone loss. The schematic pathway outlines the delineated steps by which low estrogen levels are associated with T cell activation, increased TNF-a production, and consequential augmentation of osteoclast activity, and bine resorption

Hypogonadal states can cause secondary osteoporosis. For example:

- Turner syndrome
- Klinefelter syndrome
- Kallmann syndrome
- Anorexia nervosa, malnutrition, malabsorbtion, coeliac disease, Crohn's disease, colitis, lactose intolerance, intestinal bypass surgery, or bowel resection surgery
- Liver disease and primary biliary cirrhosis
- Andropause (i.e., testosterone deficiency following surgical removal of the testes)
- Hypothalamic amenorrhea or hyperprolactinemia
- Cushing's syndrome and other deregulation of the hypothalamic-pituitary-adrenal axis
- Thyrotoxicosis, hypothyroidism
- Systemic lupus erythematosus, and associated renal insufficiency and failure
- Diabetes mellitus type 1 and 2
- Acromegaly and adrenal insufficiency, and even
- Pregnancy and lactation, where there can be a reversible bone loss
- Any condition associated with impaired calcium absorption

Furthermore, hematologic disorders linked to osteoporosis are:

- Multiple myeloma and other monoclonal gammopathies
- Lymphoma
- Leukemia
- Mastocytosis
- Hemophilia
- Sickle-cell disease
- Thalassemia

Several inherited disorders have been linked to osteoporosis. These include:

- Osteogenesis imperfecta
- Marfan syndrome
- Hemochromatosis
- Hypophosphatasia
- Glycogen storage diseases
- Homocystinuria
- Ehlers–Danlos syndrome
- Porphyria
- Menkes' syndrome
- Epidermolysis bullosa
- Gaucher's disease
- Other diseases including scoliosis, complex regional pain syndrome, Parkinson's disease, and chronic obstructive pulmonary disease

Quite surprisingly there are only few studies on RANKL or osteoprotegerin (OPG) in patients with osteoporosis. Eghbali-Fatourechi et al. (2003) determined the surface expression of RANKL on bone marrow mononuclear cells in premenopausal, early postmenopausal, and estrogen-treated postmenopausal women by flow cytometry. The surface expression of RANKL on marrow stromal cells, B cells, and T cells were significantly higher in early postmenopausal when compared to premenopausal or estrogen-treated women. These findings suggest that upregulation of RANKL on stromal cells and lymphocytes in the bone marrow could mediate increased bone resorption consecutive to estrogen deficiency. Whereas, the study by Eghbali-Fatourechi et al. (2003) refers to the early rapid phase of postmenopausal bone loss; there are data that indicate a role of the RANKL/OPG pathway in fracture susceptibility.

Research has shown an increased RANKL/OPG mRNA ratio in bone biopsies from women with hip fractures (Abdallah et al. 2005). In contrast to studies on surface expression or mRNA levels of RANKL and OPG, the measurement of these markers in serum has produced somewhat paradoxical results. With regard to OPG, most studies found elevated OPG serum levels in patients with osteoporosis (Grigrie et al. 2003), and reported decreased OPG levels in osteoporotic patients with vertebral fractures (Fahrleitner-Pammer et al. 2003).

As stated earlier (*vide supra*), bone remodeling is the process that continuously removes old material from the bone and adds new bone material. It is driven by various types of cells, most notably osteoblasts, which secrete new bone, and osteoclasts, which break bone down. The role of osteocytes is not well understood. RANKL binds to its receptor RANK, a member of the TNF receptor superfamily. RANK is expressed by preosteoclasts, and induces their conversion into mature osteoclasts. Denosumab inhibits the maturation of osteoclasts by binding to RANKL, protecting the bone from degradation and thus from osteoporosis. The drug therefore mimics the endogenous effects of OPG, another protein produced by osteoblasts which acts as an alternate receptor for RANKL, modulating the RANK/RANKL induced osteoclast activity.

The three main mechanisms by which osteoporosis develop include:

1. *Inadequate peak bone mass* (the skeleton develops insufficient mass and strength during growth).
2. Excessive bone resorption.
3. *Inadequate formation of new bone* during remodeling.

An interplay of these three mechanisms underlies the development of fragile bone tissue.

Moreover, hormonal factors strongly determine the rate of bone resorption such as:

- Lack of *estrogen* (e.g., as a result of menopause) increases bone resorption as well as decreasing the deposition of new bone that normally takes place in weight-bearing bones. The amount of estrogen needed to suppress this process is lower than that normally needed to stimulate the uterus and breast gland. The α-form of the estrogen receptor appears to be the most important in regulating bone turnover.
- *Calcium* metabolism plays a significant role in bone turnover, and deficiency of calcium and VitD leads to impaired bone deposition.
- Parathyroid glands react to low calcium levels by secreting *PTH*, which increases bone resorption to ensure sufficient calcium in the blood.
- The role of *calcitonin*, a hormone generated by the thyroid that regulates increases in bone deposition, is less clear and probably not as significant as that of estrogen.

As noted above, the activation of osteoclasts is regulated by various molecular signals, of which RANKL is one of best studied. This molecule is produced by osteoblasts and other cells (e.g., lymphocytes), and stimulates RANK. The OPG, as mentioned earlier, binds to RANKL before it has an opportunity to bind to RANK, and hence suppresses its ability to increase bone resorption. RANKL, RANK, and OPG are closely related to TNF and its receptors. The role of the wnt signaling pathway, as noted in the previous section, is recognized but less well understood. Local production of eicosanoids and interleukins is thought to participate in the regulation of bone turnover, and excess or reduced production of these mediators may underlie the development of osteoporosis (Fig. 2.2).

There are several medications used to treat osteoporosis, depending on gender. Medications themselves can be classified as antiresorptive or bone anabolic agents. Antiresorptive agents work primarily by reducing bone resorption, while bone anabolic agents build bone rather than inhibit resorption. Lifestyle changes are also important for the success of treatment, including long-term adherence to pharmacotherapy:

- *Bisphosphonates* are the main pharmacological measures for the treatment of osteoporosis, such as sodium alendronate (Fosamax) 10 mg a day or 70 mg once a week, risedronate (Actonel) 5 mg a day or 35 mg once a week, and/or ibandronate (Boniva) once a month. Data suggest that in patients who had suffered a low-impact hip fracture, annual infusion of 5 mg zoledronic acid reduced risk of any fracture by 35% (from 13.9 to 8.6%), vertebral fracture risk from 3.8 to

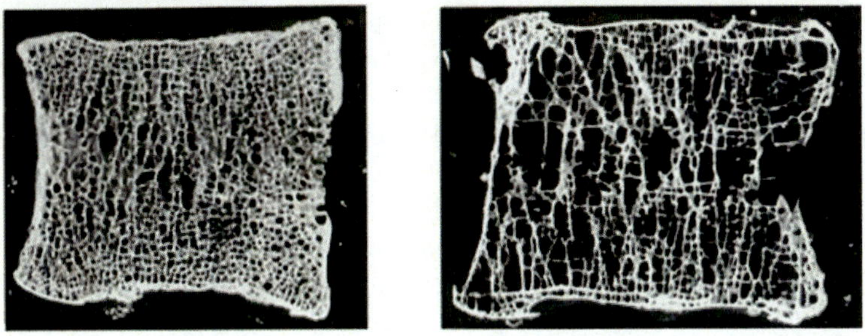

Fig. 2.2 Osteoporosis. The figure contrasts histological bone samples from a normal healthy young adult male, and from an aging adult female with evident signs of bone resorption and osteoporosis (http://www.engr.iupui.edu/~turnerch/osteoporosis.gif)

1.7%, and nonvertebral fracture risk from 10.7 to 7.6%. This study also found a mortality benefit: after 1.9 years, 9.6% of the study group (as opposed to 13.3% of the control group) had died of some other cause, indicating a mortality benefit of 28% (Lyles et al. 2007). Oral bisphosphonates are relatively poorly absorbed, and must therefore be taken on an empty stomach, with no food or drink to follow for the next 30 min. They are associated with inflammation of the esophagus (esophagitis) and are therefore sometimes poorly tolerated; weekly or monthly administration (depending on the preparation) decreases likelihood of esophagitis, and is now standard. Although intermittent dosing with the intravenous formulations such as zolendronate (zoledronic acid) avoids oral tolerance problems, these agents are implicated at higher rates in a rare but severe bone disease called osteonecrosis of the jaw. Oral bisphosphonate therapy is probably to be preferred, and doctors now recommend that any needed remedial dental work be done before treatment begins.

- *Estrogen replacement therapy* remains a good treatment for prevention of osteoporosis, but may not be recommended unless there are other indications for its use as well. There is uncertainty and controversy about whether estrogen should be recommended in women in the first decade after menopause. In hypogonadal men, testosterone appeared to yield improvement in bone quantity and quality, but studies on the effects of fractures in men with a normal testosterone level are scarce. Selective Estrogen Receptor Modulators (SERMs) are a class of medications that act on the estrogen receptors throughout the body in a selective manner. Normally, bone mineral density is tightly regulated by a balance between osteoblast and osteoclast activity in the trabecular bone. Estrogen has a major role in regulation of the bone formation–resorption equilibrium, as it stimulates osteoblast activity. Some SERMs such as raloxifene, act on the bone by slowing bone resorption by the osteoclasts. Raloxifene has the added advantage of reducing the risk of invasive breast cancer.
- *Teriparatide* (Forteo, recombinant parathyroid hormone, rPTH residues 1–34) may evince some success in osteoporosis, perhaps by acting as parathyroid

hormone to stimulate osteoblasts (*vide supra*), and increasing their activity. It is used mostly for patients with established osteoporosis (who have already fractured), and its administration is regulated by FDA, still at the time of this writing, to be used for treatment only if bisphosphonates have failed or are contraindicated. It is also widely recommended that patients with previous radiation therapy, or Paget's disease, or young patients, avoid this medication.

- *Inorganic salts*, such as calcium salts come, which as water insoluble and soluble formulations of calcium carbonate, calcium citrate, lactate, and gluconate are used to some success. Sodium fluoride treatment may cause skeletal changes such as pronounced bone density with increased number and thickness of trabeculae, cortical thickening, and partial obliteration of the medullary space, which are distinctly undesirable side-effects.
- *Alternative agents* include Denosumab, a fully human monoclonal antibody that mimics the activity of OPG, binds to RANKL, thereby preventing RANKL from interacting with RANK, and thereby reducing bone resorption. Oral strontium ranelate is an alternative "dual action bone agents" (DABAs) in that it is effective in stimulating the proliferation of osteoblasts, and inhibiting the proliferation of osteoclasts. Strontium ranelate, when taken as a 2-g oral suspension daily, can prevent vertebral and hip fracture, with notabe side effect benefits over the bisphosphonates, as it does not cause any form of upper GI complications, the most common cause for medication withdrawal in osteoporosis. It appears that strontium citrate might even be more safe and effective. In fact, the evidence to date points to the view that strontium, no matter the form, so long as water-soluble and thus ionized in the stomach acid to be protein-bound for transport from the intestinal tract and absorption into the blood stream, is effective and efficacious. Unlike drugs like sodium alendronate (i.e., Fosamax), strontium doesn't inhibit bone recycling and, in fact, may produce stronger bones.

Calcium is required to support bone growth, bone healing, and maintain bone strength and is one aspect of treatment for osteoporosis. Recommendations for calcium intake vary depending country and age; for individuals at higher risk of osteoporosis (after 50 years of age) the amount recommended by US health agencies is 1,200 g/day. Calcium supplements can be used to increase dietary intake, and absorption is optimized through taking in several small (500 mg or less) doses throughout the day. The role of calcium in preventing and treating osteoporosis is unclear – some populations with extremely low calcium intake also have extremely low rates of bone fracture, and others with high rates of calcium intake through milk and milk products have higher rates of bone fracture. Other factors, such as protein, salt and Vit D intake, exercise and exposure to sunlight, can all influence bone mineralization, making calcium intake one factor among many in the development of osteoporosis. According to the report of WHO in 2007, calcium is consumed by an acid load with food, hence it influences osteoporosis.

From the perspective of an ongoing research, one can summarize the field along with principally relevant bone-related biomarkers, which were discussed earlier (e.g., RANKL, Osteocalcin, OPG, Osteopontin, MMP-13, Calcitonin). These, and related osteoimmune markers can be reliably measured in a diagnostic protocol for osteoporosis in particular, and other osteoimmunopathologies in general (Fukunaga and Sone 2001; Kiechl et al. 2006; Lu et al. 2006; Leibbrandt and Penninger 2008), by assessing them independently or simultaneously, in only 25 μL sample volume of serum, plasma, cell culture supernatants, whole or parotid saliva, or synovial fluid (Demerjian et al. 2010). Protocols may involve SELDI-TOF or MALDI-TOF, or a FlowCytomix™ Multiple Analyte Detection System, as a bead-based immunoassay system, which is perhaps better adapted to measure cytoplasmic content and is compatible with any commercial flow cytometer. FlowCytomix™ Simplex kits are individual bead sets for the detection of one specific analyte. Combinations of Simplex Kits allow you to create your own customized analytes panel for maximum flexibility.

2.1.2.2 Osteoporosis in HIV+ and HIV/AIDS Adults

Early in the study of HIV/AIDS, observations were reported that described increased susceptibility of these patients to develop weak and fragile bones, decreased bone mineralization, and increased osteoporosis (Stephens et al. 1999; Weiel and Lenhard 2000; Guaraldi et al. 2001).

Patients with HIV-1 infection/AIDS are living longer due to the success of highly active antiretroviral therapy (HAART). In 2007, the Centers for Disease Control and Prevention in the United States reported that 16.8% of new diagnoses of HIV that year were in individuals aged over 50 years, compared to 15% in 2005. In 2005, one in four HIV-seropositive patients was older than 50 years of age. Within this decade, it is projected that over half of the HIV/AIDS population in the USA will be over the age of 50, and will suffer aging-related health threats such as cardiovascular ailments, dementia, decreased immune resilience, dyslipidemia, metabolic syndrome, diabetes, associated polypharmacy, and, pertinent to the discussion at hand, increased age-related bone disorders, such as osteoporosis, osteopenia, avascular osteonecrosis, bone fragility, and bone fracture. Thus, serious metabolic complications, including decreased bone mineral density, increased osteoporosis, and consequential raised prevalence of bone loss and fractures, are becoming alarmingly common. Management of older adults with HIV and multiple comorbidities will increasingly present challenges to infectious diseases physicians and geriatricians alike (Mondy and Tebas 2003; Simone and Appelbaum 2008; Kearney et al. 2010; Ofotokun and Weitzmann 2010).

Clinical data show that decreased bone mineral density occurs more commonly in patients with HIV than in the general population, making the group of HIV-seropositive men and women more susceptible to osteoporosis and consequential fragility fractures (Lin and Rieder 2007). Indeed, osteoporosis in HIV-infected persons is at least as prevalent as in postmenopausal women, yet this population is not listed in primary care guidelines as one that should be considered for screening (Clay et al. 2008). Whereas, the role of antiretroviral therapy (ART) in the prevalence estimates of osteopenia and osteoporosis in HIV-infected patients was initially equivocal

(Brown and Qaqish 2006), the clinical evidence linking HIV-associated osteoporosis to direct infection and ART is now clearly established (Grund et al. 2009; Arora et al. 2010). Despite these associations, however, no consistent drug-specific effect upon bone mineral density decline has been ascertained (Grund et al. 2009).

The pathogenesis or biological mechanism of these complications also remains elusive, although some research has remained enigmatic for the last decade, following it being established early on, as apparently independent from adipose tissue maldistribution, which is also common in these patients (Tebas et al. 2000). An osteoimmune hypothesis was proposed, which suggested that the proinflammatory cytokines TNF-α and IL-6, which are constitutionally produced in increased amounts in HIV-seropositive individuals, were involved because of the role we now know they play in osteoclast activation and resorption (*vide supra*) (Thomas and Doherty 2003). Moreover, data have shown significant induction of these cytokines responsible for osteoclast differentiation and bone resorption, as well as the RANKL, in peripheral blood mononuclear cells exposed in vitro to soluble HIV-1 envelope glycoprotein gp120 under experimental conditions. Moreover, pharmacologic concentrations of two protease inhibitors that are linked clinically to osteopenia, ritonavir and saquinavir, abrogate the cyto-physiological block to RANKL activity mediated by INF-γ induced degradation of the RANKL signaling adapter protein, TRAF6 (tumor necrosis factor receptor-associated protein 6) in proteasomes. By contrast, indinavir and nelfinavir, protease inhibitors that may promote or stabilize bone formation in vivo, had no effect on the RANKL system (Fakruddin and Laurence 2003). Taken together, these emerging lines of evidence give compelling support for the role of osteoimmunology in HIV/AIDS.

We have ventured the hypothesis (Barkhordarian et al. 2011) that the *ensemble* of the osteoimmune responses and pathways that converge to the observed increased bone fragility and osteoporosis and a decreased bone mineral density in HIV-seropositive patients, and that may or may not – depending of the antiviral used – be exacerbated by ART, PI, and HAART, may be consequential in large part to the immune reconstitution inflammatory syndrome (IRIS), which is often a considerable problem in the treatment of HIV-infected patients. The rationale for our hypothesis rests on the observations that IRIS is characterized by flamboyant and significant activation of both innate and adaptive immune responses with elevation of body fluid chemokines and cytokines, including common markers of inflammation such as C-reactive protein, interferon-inducible protein 10,[1] and IFN-γ and that together signify innate and adaptive immune activation (Sereti et al. 2010). IRIS+ HIV-seropositive patients show higher frequencies of effector memory T cells positive for programmed death (PD)-1, a member of the extended CD28/CTLA-4 family

[1] Also known as the chemokine, C-X-C motif chemokine 10 (CXCL10), it is produced by several cell populations (e.g., myeloid, endothelial, lymphoid) in response to IFN-γ. Recent evidence (Kwak et al. 2008) has implicated CXCL10 with a critical role in the infiltration of CD4+ T cells and F4/80+ macrophages into inflamed joints, leading to bone destruction. Since it contributes to the recruitment of inflammatory cells and is involved in bone erosion in inflamed joints, CXCL10 is an emerging key factor in osteoimmunopathological interconnections.

of T cell regulators, HLA-DR+, and Ki67+ in CD4+ T, and in regulatory T cells (Tregs, CD4+CD25+FoxP3+) cells than IRIS- HIV-seropositive patients. PD-1+CD4+ T cells in IRIS+ patients express increased levels of LAG-3, CTLA-4, and ICOS, and are driven to a TH1/TH17 cytokine profile when stimulated in vitro. Plasma levels of IRIS+ HIV-seropositive patients show marked elevation in IFN-γ and IL-7, further suggesting that IRIS is predominantly CD4-mediated phenomenon directed to reconstituting effector and regulatory T cells (Antonelli et al. 2010).

A recent modeling study defined and characterized the prevalence of and progression to low bone mineral density in HIV-infected patients by means of linear regression and logistic polytomic regression analyses of data obtained from a longitudinal cohort study of HIV-seropositive patients ($n=671$). The longitudinal analysis, aimed at identifying mixed and generalized estimating effects, relied on clinical assessments which consisted of at least one dual-energy X-ray absorptiometry scan to determine the prevalence and progression of bone mineral density, and to establish related factors. Fifty eight percent of the patients actually had more than two such scans over the course of the study (median 2.5 years of observation). The cohort had a prevalence of decreased bone mineral density of 47.5%, and osteoporosis of 23%, and clinically relevant progression to bone demineralization was observed in 28% of the patients, 12.5% to osteopenia, and 15.6% to osteoporosis, over the study period. Linear regression and logistic polytomic regression were used for the cross-sectional study and mixed effects and generalized estimating equations were used for the longitudinal study. The regression analysis identified key predictive variables:

- Age, Odds Ratio (OR): 1.07; $CI^{95}=1.05–1.08$ ($p<0.0001$).
- Gender (male), OR: 2.23; $CI^{95}=1.77–2.8$ ($p<0.0001$).
- Low body mass index (i.e., cachexia, wasting syndrome), OR: 1.14; $CI^{95}=1.11–1.17$ ($p<0.0001$).
- Duration of treatment with protease inhibitor, OR: 1.18; $CI^{95}=1.12–1.24$ ($p<0.0001$).
- Duration of treatment with tenofovir (i.e., antiviral reverse transcriptase inhibitor), OR: 1.08; $CI^{95}=1.03–1.14$ ($p<0.0019$).
- Currently in protease inhibitors treatment arm, OR: 1.64; $CI^{95}=1.35–2.04$ ($p<0.0001$).

Taken together, these statistical findings support the importance of applying adequate strategies to prevent bone demineralization and close monitoring of aging HIV-seropositive patients, specifically in certain at-risk subgroups of patients (Bonjoch et al. 2010).

To be sure, similar relationships have been noted in younger groups of HIV-seropositive patients, indicating that, whereas age is an exacerbating factor, primary HIV-1 infection is a causative factor in decreased bone mineral density. A cohort study of 33 young men (mean age ± standard deviation 38 ± 9) with primary HIV infection (mean plasma HIV RNA: 5.0 ± 1.2 $\log_{10}$ copies/mL, and matched for body mass index (22.7 ± 3.3) was followed longitudinally with regular a dual-energy X-ray absorptiometry of the lumbar spine, femoral neck and total hip, and with clinical assessments for osteopenia and osteoporosis, defined by WHO's criteria as T-scores between −1 and −2.5, and −2.5 or less, respectively. The analyses of this preliminary study demonstrated that prevalence of low bone mineral density and elevated osteoporosis in young men with primary HIV infection was associated with increased age, lower body mass index, and thyroid stimulating hormone levels, as well as with higher levels of HIV-1 viremia (Grijsen et al. 2010).

Whereas the literature largely converges on the finding that HIV-seropositivity carries with it increased risk of osteoporosis, the data in support of certain predictor variables is not absolutely homogeneous; case in point, the 2008 cross-sectional survey of 492 HIV-seropositive patients within the Aquitaine cohort. The polytomous logistic regression confirmed the predictors of older age, homosexual transmission group, low body mass index, and low HIV plasma viral load as significantly associated with the diagnosis of onset of bone abnormalities, such as decreased bone mineral density and increased osteoporosis and bone fragility in men. The analysis also showed that older age and low CD4 lymphocyte count nadir were independently associated with osteoporosis in HIV-seropositive women. In this study, the use of HAART was not found to be a significant predictor of osteoporosis after adjustment and stratification for the considered anthropometric parameters (Cazanave et al. 2008). Similarly, a study of a Slovenian HIV-seropositive cohort, while reported increased prevalence of osteoporosis among these patients in a range similar to that reported by others (i.e., prevalence among HIV-seropositive patients of prevalence of osteopenia: 47%, and of osteoporosis: 12%), ART, PI, or HAART were not found to be significant predictors for these clinical outcomes (Tomazic et al. 2007).

Alendronate appears to be a promising treatment option for HIV-seropositive patients with osteoporosis and osteopenia. Further research is required to determine the safety and efficacy of this and other available drugs. Consequently, concerted research efforts have converged toward the elucidation of bone ossification agents that may be considered for use in the treatment of osteoporosis and osteopenia in HIV-infected patients (Clay et al. 2008).

2.1.3 Ankylosing Spondylitis

Ankylosing spondylitis (Gk: ανκψλοσ, = ankylos, bent; σπονδψλοσ, = spondylos, vertebrae), was previously known as Bekhterev's disease, or Bekhterev syndrome, as well as Marie–Strümpell disease. Ankylosing spondylitis is a form of spondyloarthritis, a chronic inflammatory arthritis that can lead to fusion of the spine. Complete fusion results in a complete rigidity of the spine, a condition known as bamboo spine (Fig. 2.3).

Ankylosing spondylitis belongs to a cluster of conditions known as seronegative spondyloarthropathies, in which the characteristic pathological lesion is an inflammation of the enthesis (the insertion of tensile connective tissue into bone).

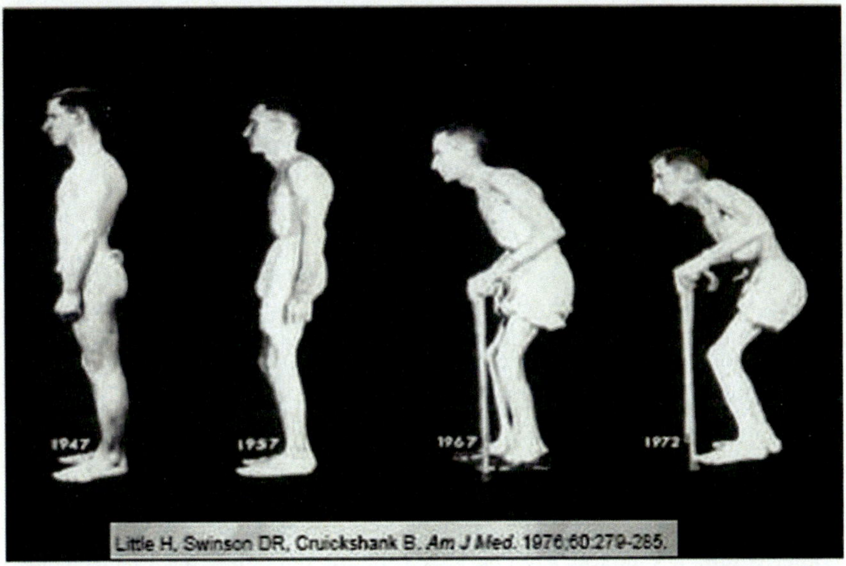

Fig. 2.3 Ankylosing spondylitis A condition that manifests as a form of chronic inflammatory osteoarthritis that can produce fusion of the spine, with consequential rigidity of the deformed spine. Adapted from: idid.us/ankylosing_spondylitis_progression.jpg&imgrefurl

Ankylosing spondylitis is a systemic rheumatic disease meaning it affects the entire body and is one of the seronegative spondyloarthropathies. Approximately, 90% of AS patients express the HLA-B27 genotype, meaning that there is a strong genetic association. However, only 5% of individuals with the HLA-B27 genotype contract the disease. HLA-B27 a class I surface antigen encoded by the B locus in the major histocompatibility complex (MHC; cf., notes) on chromosome 6 and presents antigenic peptides (derived from self and nonself antigens) to T cells. TNF-α and IL-1 are also implicated in Ankylosing spondylitis. Autoantibodies specific for Ankylosing spondylitis have not been identified. Anti-neutrophil cytoplasmic antibodies (ANCA) are associated with Ankylosing spondylitis but don't correlate with disease severity. ANCAs are a group of autoantibodies of the IgG type, which react against antigens in the cytoplasm of neutrophil granulocytes and monocytes, and detected in association with systemic vasculitis (aka, ANCA-associated vasculitides). Over 95% of people that have been diagnosed with Ankylosing spondylitis are HLA-B27 positive, although this ratio varies from population to population (only 50% of African-American patients with Ankylosing spondylitis possess HLA-B27, and it is close to 80% among patients with Ankylosing spondylitis from Mediterranean countries). In early onset disease, HLA-B7/B*2705 heterozygotes exhibited the highest risk for disease.

The association of Ankylosing spondylitis with HLA-B27 suggests that the condition involves CD8 T cells, which interact with the MHC class I, B (HLA-B). It is not proven that this interaction involves a self antigen and at least in the related

Reiter's syndrome (reactive arthritis), which follows infections, the antigens involved are likely to be derived from intracellular microorganisms. There is, however, a possibility that CD4+T cells are involved in an aberrant way, since HLA-B27 appears to have a number of unusual properties, including possibly an ability to interact with T cell receptors in association with CD4 (usually only T helper lymphocytes with CD8 reacts with HLA-B antigen as it is a MHC class 1 antigen).

The pathology mainly affects joints in the spine and the sacro-ilium in the pelvis. Ankylosing spondylitis belongs to the group of the spondyloarthropathies with a strong genetic predisposition, and a prognosis that courses quite as an autoimmune disease. The typical patient is a young adult male, aged 20–40. Symptoms of the disease first appear as chronic pain and stiffness in the lower part of the spine or sometimes the entire spine, often with pain referred to one or other buttock or the back of thigh from the sacroiliac joint.

Men are affected more than women by a ratio about of 3:1, with the disease usually taking a more painful course in men than women. In 40% of cases, Ankylosing spondylitis is associated with an inflammation of the eye (iridocyclitis and uveitis), causing redness, eye pain, vision loss, floaters, and photophobia. Another common symptom is generalized fatigue and sometimes nausea. Less commonly aortitis, apical lung fibrosis, and ectasia of the sacral nerve root sheaths may occur.

When the condition presents before the age of 18, it is relatively likely to cause pain and swelling of large limb joints, particularly the knee. In prepubescent cases, pain and swelling may also manifest in the ankles and feet, where calcaneal spurs may also develop. Pain is often severe at rest, but improves with physical activity. Concurrent inflammation and pain to varying degrees is common, regardless of rest and movement.

As noted above, there has been a longstanding claim that Ankylosing spondylitis arises from a cross-reaction between HLA-B27, and the organism's ability to counter antigens of the Klebsiella bacterial strain (i.e., nonmotile, Gram-negative, oxidase-negative, rod shaped bacteria with a prominent polysaccharide-based capsule of the Order of the *Enterobacteriales*); but, no cross reactivity with B27 has been found (and no specific antibody levels detected.

There is no direct test to diagnose Ankylosing spondylitis. A clinical examination and X-ray studies of the spine, which show characteristic spinal changes and sacroiliitis, are the major diagnostic tools. A drawback of X-ray diagnosis is that signs and symptoms of Ankylosing spondylitis have usually been established as long as 8–10 years prior to X-ray-evident changes occurring on a plain film X-ray, which means a delay of as long as 10 years before adequate therapies can be introduced. Options for earlier diagnosis are tomography and magnetic resonance imaging of the sacroiliac joints, but the reliability of these tests is still unclear. The Schober's test is a useful clinical measure of flexion of the lumbar spine performed during examination.

During acute inflammatory periods, patients with Ankylosing spondylitis will sometimes show an increase in the blood concentration of C-reactive protein and an increase in the erythrocyte sedimentation rate, but there are many with AS whose C-reactive protein and erythrocyte sedimentation rates do not increase so, normal C-reactive protein and erythrocyte sedimentation rate results do not always correspond with the amount of inflammation a person actually has. The test is not specific however, patients with Ankylosing spondylitis can also show normal level results of these biomarkers while experiencing a significant amount of inflammation in their bodies.

Variations of the HLA-B gene increase the risk of developing Ankylosing spondylitis, although it is not a diagnostic test. Those with the HLA-B27 variant are at a higher risk than the general population of developing the disorder. HLA-B27, demonstrated in a blood test, can occasionally help with diagnosis but in itself is not diagnostic of AS in a person with back pain.

In 2007, a collaborative effort by an international team of researchers in the U.K., Australia, and the United States led to the discovery of two genes, ARTS1 and IL23R, that also contribute to the cause of AS. The findings were published in the November 2007 edition of Nature Genetics, a journal that emphasizes research on the genetic basis for common and complex diseases. Together with HLA-B27, these two genes account for roughly 70% of the overall incidence of the disease.

The Bath Ankylosing Spondylitis Disease Activity Index (BASDAI), developed in Bath (UK), is an index designed to detect the inflammatory burden of active disease. The BASDAI can help to establish a diagnosis of AS in the presence of other factors such as HLA-B27 positivity, persistent buttock pain which resolves with exercise, and X-ray or MRI evident involvement of the sacroiliac joints. It can be easily calculated and accurately assesses a patient's need for additional therapy; a patient with a score of four out of a possible ten points while on adequate NSAID therapy is usually considered a good candidate for biologic therapy (van der Heijde 2004).

The Bath Ankylosing Spondylitis Functional Index (BASFI) is a functional index, which can accurately assess a patient's functional impairment due to the disease as well as improvements following therapy. The BASFI is not usually used as a diagnostic tool but rather as a tool to establish a patient's current baseline and subsequent response to therapy (Ruof and Stucki 1999; van der Heijde 2004).

No cure is known for Ankylosing spondylitis, although treatments and medications are available to reduce symptoms and pain. Normal occupations may be precluded by the symptoms of the disease.

- Physical \therapy
- Exercise
- Physical therapy and/or exercise along with *medication*

These are at the heart of therapy for Ankylosing spondylitis. Physiotherapy and physical exercises may be preceded by medical treatment in order to reduce the inflammation and pain, and are commonly followed by a physician. The goal pursued by this approach is based on the assumption that the movements will help in diminishing pain and stiffness. But, exercise per se may be contraindicated in an active inflammatory state, which could make the pain worse.

Some may require the help of walking aids such as a cane to help assist in balance and relieve some pressure on affected joints while walking and standing. Many patients with Ankylosing spondylitis find it very difficult to sit or stand for prolonged periods of time which can even be about 20 min, therefore many need to alternate times of sitting and standing, as well as times of rest. Maintaining good posture may help reduce the likelihood of a fused or curved spine, which occurs in a significant percentage of diagnosed persons.

There are three major types of medications used to treat Ankylosing spondylitis.

- *Anti-inflammatory* drugs, which include NSAIDs such as ibuprofen, phenylbutazone, indomethacin, naproxen, and COX-2 inhibitors, which reduce inflammation and pain. Opioid analgesics have also been proven by clinical evidence to be very effective in alleviating the type of chronic pain commonly experienced by those suffering from AS, especially in time-release formulations.
- *Disease-modifying antirheumatic drugs* (DMARD's) such as ciclosporin, methotrexate, sulfasalazine, and corticosteroids are used to reduce the immune system response through immunosuppression. These drugs are specifically used to reduce evidence of processes thought to underlie rheumatoid arthritis, such as a raised erythrocyte sedimentation rate, reduced hemoglobin level, raised rheumatoid factor level and more recently, raised C-reactive protein level. More recently, the term has been used to indicate a drug that reduces the rate of damage to bone and cartilage (Nandi et al. 2008).
- *TNF-α blockers*, that is TNF-α antagonists, such as etanercept, infliximab, and adalimumab (also known as biologics), are indicated for the treatment of and are effective immunosuppressants in Ankylosing spondylitis as in other autoimmune diseases. TNF-α blockers appear to be the most promising treatment, slowing the progress of Ankylosing spondylitis in the majority of clinical cases, helping many patients receive a significant reduction, though not elimination, of their inflammation and pain. They have also been shown to be highly effective in treating not only the arthritis of the joints but also the spinal arthritis associated with Ankylosing spondylitis. A drawback, besides the high cost, is the fact that these drugs increase the risk of infections. For this reason, the protocol for any of the TNF-α blockers includes a test for tuberculosis (like Mantoux or Heaf) before starting treatment. In case of recurrent infections, even recurrent sore throats, the therapy may be suspended because of the involved immunosuppression. Patients taking the TNFα medications are advised to limit their exposure to others who are or may be carrying a virus (such as a cold or influenza) or who may have a bacterial or fungal infection.

Unattended cases of Ankylosing spondylitis that are accompanied by dactylitis or enthesitis, especially when spine inflammation is not yet active, may result in a misdiagnosis of normal rheumatism. In a long-term undiagnosed period, osteopenia or osteoporosis of the Ankylosing spondylitis spine may occur, causing eventual compression fractures and a back "hump." Typical signs of progressed Ankylosing spondylitis are the visible formation of syndesmophytes on X-rays and abnormal bone outgrowths similar to osteophytes affecting the spine. The fusion of the vertebrae paresthesia is a complication due to the inflammation of the tissue surrounding nerves.

Organs commonly affected by Ankylosing spondylitis, other than the axial spine and other joints, are the heart, lungs, eyes, colon, and kidneys. Other complications are aortic regurgitation, Achilles tendinitis, atrio-ventricular node block, and amyloidosis. Owing to lung fibrosis, chest X-rays may show apical fibrosis while pulmonary function testing may reveal a restrictive lung defect.

2.2 Bone Tumors and Necroses

Bone tumors are abnormal growth of cells within the bone. They may be noncancerous (i.e., benign) or cancerous (i.e., malignant). For example, osteochondromas are the most common noncancerous bone tumors, and occur most often in people during the growth spur between the ages of 10 and 20.

Although some bone tumors may remain asymptomatic for a relatively long period of time – up to years – due to their slow progression, most tumors of the bone present a rather characteristic panorama of symptomatology, which can be briefly outlined as:

- Relative ease of bone fracture following slight injury or trauma
- Pain, which increases at rest and at night
- Occasionally, a mass or a swelling at the sensitive site

Taken together, these manifestations call for a more in-depth and aggressive diagnostic exploration, which will involve X-rays, bone biopsies, as well as bone scans and imaging studies (magnetic resonance imaging, MRI).

2.2.1 Osteochondroma

Osteochondromas typically involve the bone and the cartilage tissues, and present near the end of a long bone, as benign capped-cartilage outgrowths, connected to bone by a stalk. They account for approximately 35% of benign bone tumors, and 9% of all bone tumors, and are commonly referred to, in lay language, as a "bone spur," and in academic circles as "osteocartilaginous exostosis."

Exostosis is an osteochondropathy (i.e., disease of the bone and cartilage) that involves the process of formation of new bone on the surface of existing bone. It is most often accompanied by a sensation of pain that may range from mild and dull to sharp, constant, and debilitatingly severe pain. Fewer than 1% of osteochondromas may progress to malignancy of the cartilage cap resulting in secondary chondrosarcoma. Hereditary multiple exostosis is an autosomal dominant condition associated with short stature, multiple osteochondromas, and asymmetric growth at the knees and ankles, which may lead to deformities, and there is evidence that mutations in these two genes are responsible for over 70% of the multiple exostosis cases. Although at the time of this writing the etiology of osteochondromas remains to be elucidated, molecular analyses of the Exostosin (EXT)-1 and EXT2 genes indicate that mutations at these sites lead to multiple osteochondroma (Heinritz et al. 2009). It is noteworthy to note that all members of this multigene family encode glycosyltransferases involved in the adhesion and/or polymerization of heparin sulfate chains at heparin sulfate proteoglycans, which play a key role in regulating chondrocyte proliferation and differentiation (Jennes et al. 2009). The process of osteochondroma formation may originate from a herniation of a peripheral portion of the physis, possibly due to an idiopathic cause, or resulting from trauma, or perhaps even a perichondrial ring deficiency. At the site of the herniation, an abnormal extension of metaplastic cartilage responds to growth factors, which yields to exostosis by means of a zone of an initial lesion zone (osteochondroma, Stage I), followed by cell proliferation, osteon columniation, hypertrophy, calcification, ossification (osteochondroma, Stage II), and aggressive local invasion (osteochondroma, Stage III) (Bovée 2008).

2.2.2 Bone Malignancies

The principal malignant bone cancers include:

- *Chondrosarcoma*, a cartilage-based tumor that represents about 25% of primary bone cancers. From an oestoimmune perspective, it is interesting to note that immune factors seem to be intimately involved in the regulation of growth of these tumors as recent research has revealed that ADAMTS9, an important member of the disintegrin and metalloproteinase with thrombospondin motifs (ADAMTS) gene family, is strongly activated by IL-1β in both chondrosarcoma cells and human chondrocytes, via the molecular pathway recognized as the cytoplasmic 1 protein of the Nuclear Factor of Activated T cells (NFAT) family, NFATc1 (Northrop et al. 1994; Yaykasli et al. 2009), which may not be surprising when one considers that the transcriptional activity of NFATc1 is enhanced by the Pim-1 kinase (Rainio et al. 2002), itself a regulator of hematopoietic cell growth and differentiation (Hammerman et al. 2005), and a proto-oncogene implicated in a variety of human cancers, albeit not detectable in the HCS-2/8 stable human chondrosarcoma cell line by the methodologies available in the 1990s (Zhu et al. 1994). This ninth member of the ADAMTS family is endowed with aggrecan-degrading activity; aggrecan (aka, large aggregating proteoglycan, chondroitin sulfate proteoglycan-1) being, along with Type-II collagen, a major structural component of cartilage.
- *Fibrosarcoma*, a malignant tumor of the fibrous connective tissue and characterized by immature proliferating fibroblasts and undifferentiated anaplastic spindle cells that can also involve the periosteum and the overlying muscle (approximately 5% of all primary bone sarcomas), whose cells produce neither bone nor cartilage, but collagen primarily. Bone lesions of fibrosarcoma are generally osteolytic, and osteoclast-mediated. Histiocytes, that is cells of the immune system located within connective tissue that engulf cellular debris and

pathogens, are also found (e.g., CD68+ monocytes/macrophages). In the osteoimmune context of the present discussion, it is important to note that fibrosarcoma cell-derived IL-1α has been noted to induce strongly cell-mediated immunity in situ in a murine model, whereas fibrosarcoma cell-derived IL-1β counters this immune response (Marhaba et al. 2008). Indeed, the proinflammatory cytokine IL-1β strongly promotes tumor growth by inducing myeloid-derived suppressor cell and regulatory T cell (Treg, vide supra) expansion. It follows that IL-1α could efficiently support a fibrosarcoma tumor vaccination protocol, so long as myeloid-derived suppressor cell and Treg expansion mediated by IL-1β is controlled by means other than the IL-1 receptor inhibitor, which is the preferential target of IL-1α (Weiss et al. 2009).

- *Ewing's sarcoma,* named after the American pathologist James Ewing (1866–1943), is a relatively rare form of round-cell bone cancer, which affects the pelvis, the femur, the humerus, and the ribs predominantly of male teens and young adults, and which may shed cancer cells into surrounding soft tissues (Ewing 1919). Osseous Ewing sarcomas are thought to derive from a bone marrow of mesenchymal cell origin. Biopsies are typically positive for CD99, a marker typically found on immature T cell precursors (e.g., double negative CD4/CD8 thymocytes), but also involved in lipid-raft mobilization and concomitant enhancement of the earliest T cell receptor (TcR) signaling events that engage tyrosine phosphorylation of the TcRz chain and consequent enhancement of TcRz-mediated signal 1 (Oh et al. 2007). Biopsies are also negative for the common leukocyte antigen, CD45, which suggests that the tumor may be evading surveilling immune cell populations (Bernstein et al. 2006). Taken together, this evidence, and the observation that patients with metastized Ewing sarcomas have significantly elevated proprotions and absolute numbers of immunosuppressive CD4+CD25+FoxP3+ regulatory T cells (Tregs) (Brinkrolf et al. 2009), confirm an important role of cellular immunity in the metabolism of this bone malignancy.
- *Osteosarcoma,* is the most aggressive, and most common cancerous neoplasm arising from primitive transformed cells of mesenchymal origin that exhibit osteoblastic differentiation and produce malignant osteoid. It is the sixth leading cancer among adolescents, and afflicts 4–5% of the young adult population. It is also common among the aging population, and the aged. It is predominant in the metaphyseal region of tubular long bones (e.g., femur [60–70%], tibia, radius), although it also manifests in the maxilla and the mandible [7–10%]. The lesions are characterized by the presence of osteoid bone formation with or without calcification, pleomorphic actively growing cells, multinucleated osteoclast-like giant cells. The tumor can present as:
  - *Stage I* – rare, but including parosteal osteosarcoma or low-grade central osteosarcoma (excellent prognosis (>90%) with wide resection)
  - *Stage II* – determined by the site of the tumor (proximal tibia, femur, pelvis, etc.), the size of the tumor mass (in cm.), and the degree of necrosis from neoadjuvant chemotherapy (chemotherapy prior to surgery). Expressions of

p-glycoprotein, CXCR4-positive, or Her2-positive are predictive of distant metastases to the lung (prognosis dependent upon the time to metastasis, the locus of metastasis, and the extent of metastasis)
- *Stage III* – lung metastates (prognosis is bleak [30% at best], and depends on the resectability of the primary tumor and lung nodules, the degree of necrosis of the primary tumor, and the number of metastases)

The incidence of osteosarcoma, as other bone cancers is higher in families with certain familial cancer syndromes. We have reported that the microenvironment of osteosarcomas is invaded by lymphocytes, and regulated by neuroendocrine-immune feedback loops (Angeli et al. 2002; Prolo et al. 2003). Osteosarcoma-invading lymphocytes exhibit signs of activation. The immune processes that are engaged within the malignant bone matrix involve the production of cytokines, which regulate the process of apoptotic programmed cell death. We explored the mechanisms by which apoptosis of osteosarcoma cells might be modulated by the neuroendocrine-immune system, and potential physiological implications (Prolo et al. 2003). Taken together, this evidence has yielded promising novel and cutting-edge avenues of immune interventions directed to this bone malignancy (Marina and Gorlick 2009).

In closing, it is important to note that, due in part to the lymphatic and blood circulatory supply to the bone, several malignancies metastize to bone. Such metastases, while invading and involving the bone tissue, are not considered proper bone tumors. Multiple myeloma, for instance, often affects the bone, but is not considered a bone tumor, and so breast, kidney, lung, prostate, and thyroid cancers, which spread to the bone, particularly among the elderly, but are not bone cancers, per se.

2.2.3 Osteonecrosis

Osteonecrosis has two distinct facets, which must not be overlooked: first, it must be understood that necrosis (and rebuilding) of bone is a normal physiological event; old bone is continuously reabsorbed and replaced with new bone, a process that ensures that the skeleton is strong and vigorous, and that an adequate balance of minerals is always maintained under check. Second, when the replacement of reabsorbed fails to occur, then a pathology ensues, that is characteristically referred to as osteonecrosis.

In brief, osteonecrosis, when not counterbalanced by new bone synthesis, is a disease. One among the principal reasons why new bone synthesis may not sustain the rate of bone resorption is often found in physiopathological conditions that block, blunt, hinder, or prevent bone synthesis, while simultaneously baring the survival of existing bone components. For example, osteonecrosis occurs in situations where there is cellular necrosis intrinsic to the bone components due to interruption of the blood supply, and consequential deprivation of oxygen. Collapse of the bone structure ensues. A number of etiological causes have been presented, ranging from alcoholism, anemia, embolism, thrombosis, wanton abuse of steroids, hypertension, radiation and bisphosphonates, particularly in the case of ostenecrosis of the mandible. An idiopathic etiology (no cause is found) is also possible.

Osteonecrosis is a sequential process, which recapitulates, in a certain sense, the very osteoimmune milieu we have been describing. Upon onset of the osteonecrotic process following anoxia, the hematopoietic cells, which are precursors to the blood cells, including the cells of the immune system, are the first cells to undergo necrosis, usually within 10–12 h. The bone cell populations we have discussed above, such as osteocytes, osteoclasts, and osteoblasts are more resistant to anoxia, and subsequently enter into irreversible necrosis. Within a 48-h span, these populations too have died. The bone marrow fat cells are the most resistant, and survive anoxia for up to 60–120 h.

Perfusion of oxygenated blood can arrest and repair osteonecrotic bone, provided the damage is not advanced. Repair of ischemic bone occurs first by means of an angiogenic process that permits the movement of undifferentiated mesenchymal cells from adjacent living bone tissue that grow into the dead marrow spaces, and, most importantly, access by macrophages and other phagocytic cells to degrade cellular debris. Healing of osteonecrotic tissue then proceeds to cellular differentiation of the mesenchymal cells into osteoblasts or fibroblasts. The remaining inorganic mineral contributes to the formation of the necessary framework for the establishment of new, fully functional bone tissue.

Related, but distinct to osteonecrosis, is the clinical condition characterized, not so much by necrotic death of the cellular components of bone, but by softening of the bones due to defective bone mineralization, consequential to Vit D deficiency, and any metabolic condition that may result in a defect in mineralization of the osteoid protein framework, a condition known as osteomalacia (Hu et al. 2010).

2.3 Stomatological Osteoimmunopathology

2.3.1 Osteonecrosis of the Jaw

It is probable that the first modern scholarly description of the pathology of osteonecrosis of the jaw can be traced to Bond's treatise (Bond 1848) on dental medicine. It then was described with localized or generalized deep ache or pain, often initially of multiple mandibular sites, but did not manifest in individuals with good gingival health and apparently sound teeth and no abscess or suppuration. Today, it is clear that mandibular and maxillary osteonecrosis involve a pathological process in the cancellous bone and bone marrow that results from bone infarcts and damage to the deeper portion of the cancellous bone mediated by a range of local and systemic factors. Cancellous bone is more prone to this pathology than cortical bone, because of its rather loose mesh-like structure, with spaces characteristically filled with marrow tissue that is prone to hypoxia consequential to infarcts deep in the tissue and leading to premature death by indiscriminate necrosis[2] of the cells resident in the osteo matrix. Cell loss within the bone even in its mild or minor forms, creates an environment[3] that is conducive to bacterial growth, local infection, and inflammation. In other words, while osteonecrosis of the jaw is not, per se, an infection or an infectious disease, it generally carries secondary low-level or severe bacterial infection and chronic nonsuppurative osteomyelitis.

Necrosis of the facial skeleton can be particularly painful due to trigeminal neuralgia associated with the infection/inflammatory event as described earlier. So much so that it has earned a name for itself in the stomatological pathology literature as Neuralgia-inducing cavitational osteonecrosis. A typical pathological description of the condition reports a nutrient-starved necrotic bone or bone marrow with manifestation of a greasy, dead fatty marrow (wet rot), a very dry, sometimes leathery marrow (dry rot), or a completely hollow marrow space (osteocavitation).

Whereas it is not the focus of this writing to describe the etiology and pathology of this condition, it must be emphasized that a definitive symptom is the exposure of mandibular or maxillary bone through lesions in the gingiva that do not heal. Lesions may be consequential to dental work, tooth extraction, or other forms of trauma. Toxic agents, such as heavy metals (e.g., cadmium), nicotine, and acetaldehyde derived from alcohol metabolism have also been implicated as etiological agents for mandibular and maxillary osteonecrosis. Oral bisphosphonates,[4] such as zoledronic acid, pamidronate and alendronate, risedronate, and ibandronate, increase the risk for osteonecrosis, perhaps interfering with matrix metalloproteinase-2, a zinc-dependent endopeptidase, which plays a critical role in cell-mediated immune surveillance (e.g., immune cell activation, proliferation, migration, differentiation), as well as other cell types.

Data suggest that nitrogen-containing bisphosphonates inhibit cytoplasmic farnesyl diphosphate synthase,[5] and prevent prenylation of essential small GTPase signaling proteins. Thus, blunting the activity of farnesyl diphosphate synthase is associated with antitumor effects, as well as with the activation of gd-T cells, small subset of T cells[6] found abundantly in the mucosal immune system and that seem to have a prominent role in the recognition of lipid antigens (Roelofs et al. 2006). Emerging evidence further indicates that in addition to their effects on bone pathology, bisphosphonates are endowed with anticancer activity. These effects

---

[2] Note: indiscriminate necrosis signifies that osteoclasts also die by necrosis; thus, necrotic bone does not show signs of bone resorption.

[3] Anatomists talk about "danger spaces," that is spaces either in normal or in pathological anatomy that bacteria can invade and where they can expand with no or little immune surveillnce.

[4] Biphosphanates are drugs with two phosphonate groups (PO_3) that are used to prevent the loss of bone mass in osteoporosis and similar diseases, in that they inhibit bone resorption by osteoclasts.

[5] An enzyme of the mevalonate pathway.

[6] This subset has been referred to as the "first line of defense," "regulatory cells," or "bridge between innate and adaptive responses," because the can both act as T cells (i.e., rearrange TCR genes to produce junctional diversity and will develop a memory phenotype), and innate immune cells (Holtmeier and Kabelitz 2005).

appear to be mediated via the osteoimmune TNF-α pathway, and to involve the RANKL/OPG axis (Green and Clézardin 2010).

2.3.2 *Temporomandibular Disorders*

The term "temporomandibular joint disorder (TMD)" is a generic term used to refer to a syndrome of TMJ malfunctions and symptomatologies, which arise from acute or chronic inflammation of the TMJ. That is to say, TMD is the osteoimmune pathology *par excellence* in the head and neck region generally speaking and within the anatomical region defined as the stoma in particular.

The TMJ is susceptible to many of the conditions, which were discussed above, that bridge the domains of osteology and immunology, and that therefore affect any other joints in the body, including ankylosis, arthritis, trauma, dislocations, developmental anomalies, and neoplasia. Moreover, the TMJ syndrome and resultant dysfunction can result in significant pain and functional impairment. In addition, and because of the intimate anatomical relationships within the domain of the TMJ, which were discussed earlier (*vide supra*), the TMJ disorder complex transcends the boundaries between dentistry, immunopathology, physiology, clinical anatomy, and neurology, to cite a few.

Signs and symptoms of the TMJ syndrome disorder can vary in their presentation, and severity; they can be very complex or exceedingly simple to recognize and to interpret. They may involve one or several components of the TMJ:

- *Muscles of mastication*: There are four paired muscles of mastication, which are all innervated by branches of the anterior/motor root the mandibular branch (V3) of the trigeminal nerve, and are supplied by branches of the maxillary artery, because of their shared embryological origin from the first branchial arch. These muscles originate on the skull and insert into the mandible, thereby allowing for jaw movements during contraction. While accessory muscles (e.g., tongue) participate in the several functions of the jaw, the four primary muscles of mastication are:
  - The *masseter* for the elevation and protraction of the mandible, and antagonized by the platysma muscle: a thick, somewhat quadrilateral muscle, consisting of two parts: the *larger superficial portion*, which arises as a thick, tendinous aponeurosis from the zygomatic process of the maxilla, and from the anterior two-thirds of the lower border of the zygomatic arch, and whose fibers course downward and backward, to be inserted into the angle and lower half of the lateral surface of the ramus mandible, and the *smaller and deep portion*, which arises from the posterior third of the lower border and from the entire medial surface of the zygomatic arch, and its fibers course downward and forward, to be inserted into the upper half of the ramus and the lateral surface of the coronoid process of the mandible. The deep aspect of the masseteric muscle lies medial to, and is thus concealed in a frontal approach, by portions of the superficial portion anteriorly, and the parotid

gland posteriorly. It is supplied by the masseteric artery, the second inferior branch of the middle maxillary artery, as it runs laterally through the mandibular notch to the deep surface of the muscle.

- The *temporalis* for the elevation and retraction of the mandible and antagonized by the platysma muscle: the muscle that can be seen on the temples and felt contracting while the jaw is clenching and unclenching, and which is covered by the temporal fascia (i.e., temporal aponeurosis), arises from the temporal fossa and the deep part of temporal fascia, passes medial to the zygomatic arch, and inserts onto the coronoid process of the mandible. It is supplied by the anterior and the posterior deep temporal arteries, superior branches of the middle maxillary artery, and is innervated by the deep temporal nerves that are branches from V3. Retraction of the mandible is brought about by the horizontal fibers in the posterior aspect of the temporalis. The *sphenomandibularis* fibers attach to the sphenoid bone and the mandible, as integral part of the temporalis.[7]
- The *medial (internal) pterygoid* for elevation and protrusion of the mandible, closure of the jaw, and participation in side motion of the jaw: a thick, quadrilateral muscle that consists of two heads: the *deep head* from the medial side of lateral pterygoid plate behind the upper teeth, and the *superficial head* from the pyramidal process of palatine bone and maxillary tuberosity. Together the fibers run inferiorly, laterally, and posteriorly to insert strong tendinous lamina into the infero-posterior aspect of the medial surface of the ramus and angle of the mandible, as high as the mandibular foramen. The medial pterygoid is innervated by the medial pterygoid branch of the anterior/motor root of the mandibular branch of the trigeminal nerve. This nerve is important in the deep face, not only because it also innervates the tensor tympani in the tympanic cavity, and the tensor veli palatini, the only of the five paired skeletal muscles to the soft palate muscle not innervated by the pharyngeal plexus, but also because fibers of the medial pterygoid nerve project to the parasympathetic *otic ganglion*[8] located inferior to the foramen ovale in the infratemporal fossa. The muscle is supplied by an irregular number of pterygoid branches of the middle portion of the maxillary artery.

---

[7]Those fibers were reported in the 1990s, and reported as a putative fifth muscle of mastication (Dunn et al. 1996). This scientific communication was retrieved in 2008, as it had become clear that the fibers were part of the temporalis.

[8]The otic ganglion projects sympathetic postganglionic fibers that contribute to the plexus surrounding the middle meningeal artery. Preganglionic parasympathetic fibers reach the otic ganglio via the glossopharyngeal nerve, and to a smaller extent via the facial nerve, as the lesser petrosal nerve continued from the tympanic plexus. The ganglion–postganglionic parasympathetic fibers course with sympathetic fibers mainly in the auriculotemporal nerve to supply the parotid gland (secretomotor).

N.B.: A slender filament (sphenoidal) ascends from it to the nerve of the Pterygoid canal, and a small branch connects it with the chorda tympani, and is connected by two or three short filaments with the nerve to the Pterygoideus internus, from which it may obtain a motor, and possibly a sensory root.

– The *lateral (external) pterygoid* for depression and protrusion, and side-to-side motion of the mandible: the sole muscle whose function is to open the jaw through protraction of the condylar processes along the articular eminences; the beginning of this action is aided by the accessory muscles of mastication, the digastric, mylohyoid, and geniohyoid muscles. The lateral pterygoid also consists of two heads: the *superior head* originates on the infratemporal surface and infratemporal crest of the greater wing of the sphenoid bone, and inserts onto the articular disc and fibrous capsule of the temporo-mandibiular joint. The *inferior head* arises from the lateral surface of the lateral pterygoid plate, and inserts onto the pterygoid fovea under the condylar process of the mandible. Innervation is provided by the lateral pterygoid nerve (aka, external pterygoid nerve), a branch of the maxillary (V2) or separately from the anterior division of the mandibular (V3) nerves, in conjunction with the nerve to the buccinator.[9] The same pterygoid branches of the middle maxillary artery supply the internal and the external pterygoid muscles.

- *Innervation*[10]: As noted above, branches of the trigeminal, and projection to related ganglion within the deep face or the central nervous system – including rather distally related physiological correlates (e.g., vagal reflex, nausea) contribute to the complex innervation pattern of the TMJ. Sensory and motor innervation, as well fibers projecting to and from the autonomic ganglia, with sympathetic and parasympathetic components, highlight the complexity of the normal physiology of the TMJ, as well as the multifaceted proximal and distal pathological *sequelae*[11] related to TMJ dysfunction, from impaired speech and

[9]The paired buccinators muscles are accessory muscles of mastication because they pull back the angle of the mouth and to flatten the cheek area, which aids in holding the cheek to the teeth during chewing. The fibers originate anteriorly from the outer surfaces of the alveolar processes of the maxilla and mandible at the three molar teeth, and posteriorly, from the anterior border of the pterygomandibular raphé that separates it from the constrictor pharyngis superior. The fibers course to converge toward the angle of the mouth and intersect each other with the orbicularis oris. Motor innervation is from the buccal branch of the facial nerve (VII), and sensory innervation is from the buccal branch of V3. It is supplied by the buccal artery, the last branch of the middle maxillary artery.

[10]Inflammation at the TMJ proximal to the branches of the trigeminal (e.g., auriculotemporal nerve) may produce swelling and compression of said nerve, with consequential damage to the trigeminal nerve, constant or acute pain even without movement of the jaw, as well as a myriad of physiopathological manifestations (vide infra).

[11]Often mediated by retro-grade transport signaling to the gasserian nucleus (aka, trigeminal ganglion, gasserian ganglion), the sensory ganglion of the trigeminal nerve (CN V) in the Meckel cavity within the dura mater that covers the trigeminal impression near the apex of the petrous part of the temporal bone. Its motor root courses antero-medial to the sensory root, and inferior to the ganglion, thus leaving the skull through the foramen ovale; immediately below this foramen, it joins the mandibular nerve. From the ganglion convex border, three large nerves proceed, the ophthalmic (V1), maxillary (V2), and mandibular (V3) trigeminal nerves, of which V1 and V2 are exclusively sensory, and V3 is joined outside the cranium by the motor root. Animal research has shown that the trigeminal ganglion projects to trigeminal brain stem areas (principalis, spinal trigeminal nucleus, interpolaris, and caudalis), which supports the possibility of far-ranging systemic complications of TMJ dysfunction (Demerjian et al. 2010).

mastication, to tenderness and pain at the joint, to migraine, muscle spasms, dyskynesia and dystonia, and others.

- *Ligaments*: The *temporomandibular ligament* is the major ligament of the TMJ, and consists of the thickened lateral aspect of the capsule, with a distinct outer oblique portion, and an inner horizontal portion. Its primary function is to define the farthest extents of movements, of the mandible minor ligaments that are accessory to the TMJ include:
  - The *stylomandibular ligament* (styloid process to the angle of the mandible), which separates the infratemporal region anteriorly from the parotid region posteriorly.
  - The *sphenomandibular ligament* (spine of the sphenoid bone to the lingula of mandible).
  - The *oto-mandibular ligaments* (i.e., disco-mallear[12] and malleo-mandibular ligaments) that conjoin the malleaus bone of the middle ear with temporomandibular joint.
- *Bones*: clearly one among the principal etiologic factors of TMJ disorders involve degenerative joint inflammation-induced disease, such as osteoarthritis or organic degeneration of the articular surfaces, recurrent fibrous and/or bony ankylosis, tumor, bone necrosis, or developmental abnormality, as discussed in this writing.
- *Teeth*: Impaired tooth mobility and tooth loss can be caused by destruction of the supporting bone (e.g., often consequential to osteoporosis) and by heavy forces being placed on teeth, such as that applied by involuntary clenching and grinding[13] during psycho-emotional stress. The movement of the teeth affects how they contact one another when the mouth closes; the equilibration of forces of mastication and therefore the displacements of the condyle, and the overall relationship between the teeth, muscles, and joints can be altered in TMJ disorders. Inflammation of the inner core of the tooth (aka, pulpitis) also may result from excessive surface erosion due to grinding and bruxism. Often TMJ disorders may be consequential to dental treatment that required extraction of the wisdom teeth or back molars, thus precluding adequate support of the teeth onto each other upon closing the mouth, and consequential strain on the joint ligament and musculature. Such situations are not uncommon in edentulous patients whose dentures must be finely adjusted to provide support. A similar situation occurs generally in the aging middle-age population, as normal wear and tear of the teeth surface, and age-related bone resorption (*vide supra*) may contribute to a

---

[12] Described within the last 15 years: Rodríguez-Vázquez et al. (1998); Rowicki and Zakrzewska (2006).

[13] i.e., Bruxisms – an oral parafunctional diurnal and noctural activity that occurs in most humans at some time in their lives, and that may, when it occurs during sleep, lead to major of health issues, even if it occurs during short naps, from common sleep disorders, to sleep apnea and consequential cardiovascular ailment, possibly *dementia praecox*, parkinsonian-like symptoms, dystonia and dyskenesia, etc.

shortening of the vertical dimension of the aperture of the jaw, and overly tense neuromuscular activity of the TMJ structures as outlined earlier (*vide supra*)
- *Ear*: balance, auditory sensations, including hearing loss or tinnitus, the perception of sound within the human ear in the absence of corresponding external sound.

The symptomatology of TMJ disorder might invoke manifestations such as:

- Biting or chewing difficulty, discomfort or pain
- Clicking, popping, or grating sound[14] when opening or closing the mouth, impaired movement of the jaw (e.g., difficulty in closing the mouth)
- Dull, aching pain in the jaw, face, ear (i.e., otalgia), neck or shoulders (e.g., fibromyalgia, myofascial pain dysfunction syndrome)
- Morning headache and migraines (particularly in the morning)

These and other neurological disorders, including dystonias, and Parkinson's disease, which are common in aging adults, may find a common etiology in TMD. One possible mechanism involves the innervation about the articulation of the joint itself. Demerjian et al. (2010) proposed an interesting model, by which the auriculotemporal nerve, a branch of the mandibular nerve, which innervates the temporomandibular joint and courses to the tympanic membrane and anterior cochlear surface and neighboring tissues, may play a critical role. Auriculotemporal nerve fibers project to the sympathetic otic ganglion. Clenching, grinding, trauma, bone loss, and stress can change the jaw bite, and decrease its vertical dimension. Subsequent irritation and compression of the auriculotemporal nerve can occur, with associated paresthesia, pain, and discomfort. Symptoms can be local and specific (e.g., TMD), as well as varied and systemic (e.g., neurologic, dystonic and neuromuscular disorders, including tremors, muscle spasms leading to impaired and awkward positional control of the head, hands, other extremities, speech impairment, incontinence, impaired sleep, associated depressive symptomatology) (Demerjian et al. 2010).

Clinical intervention, aimed at changing the maxillomandibular occlusal relationship by changing the vertical dimension of the patient's jaw bite, may relieve auriculotemporal compression and associated irritation, and lead to resolution of the clinical symptoms (Demerjian et al. 2010).

Radiographic, arthrotomographic, and histologic data show the anatomic relationship between the joint components and the different nerves running in the vicinity of the joint. The intimate anatomical relationship of the auriculotemporal nerve and the superficial branches of the temporal artery contributes to the source of migraine headaches and other referred pains in certain patients with TMD (Demerjian et al. 2010).

---

[14] Note: This is characteristic of TMJ disorders, and is due to a dysfunctional relationship between the condyle of the mandible and the articular disc. The sounds produced by this dysfunction are termed "clicks" or "pops" when a single sound is heard, and "crepitation" or *crepitus* (Lt., rattle, crackilng) when sounds are multiple or rough in nature.

At the posterior border of the lateral pterygoid muscle, the nerve trunk is in direct contact with the condylar neck, thus being at risk of injury as occlusion of the temporomandibular joint is altered by bone loss, trauma, grinding, and other pathological etiologies. Indeed, in most cases, the auriculotemporal nerve may be close to 10.0–1.5 cm inferior to the superior surface of the condyle and a few mm posterior to the neck of the condyle, thus in an anatomical location that position that is at high risk for entrapment with adjacent tissues. When the auriculotemporal nerve courses between the condyle and the temporal bone glenoid fossa wall, the risk of mechanical compression during medial synovial disc displacement is increased, and with it the associated pain with temporomandibular joint movements (Demerjian et al. 2010).

Additional disaggreable sensations may project to the terminal area of distribution of the nerve branches in the vicinity of the joint, ranging from the ear, temple, cheek, tongue, and teeth, and beyond the head and neck proper. Systemic exacerbations of pathologies of the AT branch of the trigeminal nerve typically arise as manifestations of the trigeminal reflex, which involves the Gasserian nucleus (or ganglion):

- The *trigemino-cardiac reflex* presents as parasympathetic dysrhythmia, sympathetic hypotension, apnea, or gastric hypermotility, consequential to the stimulation of any of the sensory branches of the trigeminal nerve.
- Involvement of the *Gasserian ganglion* leads to the mechanical stimulation of all the central and peripheral branches of the trigeminal nerve.
- Patients may develop severe bradycardia, asystole, and arterial hypotension, and other *cardiovascular symptoms*, which subside almost immediately with cessation of the stimulus.
- Patients may also manifest a spectrum of *neurologic disorders* as well as multisynaptic neck muscle withdrawal responses, mediated by neural circuits at brainstem level, and absent or impaired in patients with Parkinson's disease or progressive supranuclear palsy, putatively because of a degeneration of brainstem neural structures.

2.3.3 *Fanconi Anemia*

Significant alterations in the bone marrow are among the clinical manifestations of note in Fanconi[15] anemia. This condition results from a genetic defect in a cluster of proteins responsible for DNA repair, and thirteen genes have been identified whose mutations lead to Fanconi anemia: FANCA, FANCB, FANCC, FANCD1,

[15] The condition, which now carries his name, was originally described as hereditary panmyelopathy with short stature and hyperpigmentation by the Swiss pediatrician Guido Fanconi (1892–1979). Whereas the incidence of the disease is relatively rare (1/350,000), Ashkenazi Jews (1/90) and Afrikaners in South Africa are at high risk.

FANCD2, FANCE, FANCF, FANCG, FANCI, FANCJ, FANCL, FANCM and FANCN. At least eight of these proteins, FANCA, -B, -C, -E, -F, -G, -L, and -M, assemble to form a core protein complex in the nucleus. Assembly is activated by replicative stress, DNA damage, or reactive oxygen species. FANCA and FANCG multimerize when the cell experiences such oxidative stress-induced damage. Following assembly, the protein core complex activates FANCL protein, an E3 ubiquitin-ligase that monoubiquitinates FANCD2 into FANCD2-L, which interacts with the BRCA1/BRCA2 complex. Alternatively, through ionizing radiation, FANCD2 can be phosphorylated by protein complex ATM/ATR, thus altering S-phase checkpoint control (Taniguchi and Dandrea 2002; Yuan et al. 2010).

Patients with Fanconi anemia are at increased risk for cancer, and more specifically acute myelogenous leukemia.[16] Nine out of ten patients with Fanconi anemia develop bone marrow failure, and consequentially manifest the inability to produce blood cells from hematopoietic precursors. Patients also suffer from a spectrum of congenital defects, commonly short stature (i.e., impaired bone growth), abnormalities of the skin, arms, head, eyes, kidneys, and ears, and other developmental disabilities (Yamashita and Nakahata 2001).

The expression of the transcription factor Fanconi anemia zinc finger protein (FAZF[17]), closely related to the transcription factor promyelocytic leukemia zinc finger protein (PLZF), itself composed of an N-terminal BTB/POZ and C-terminal zinc finger motifs, as an upstream factor of CBFA1 (Runx2/core-binding factor-1),[18] is increased by BMP-2 in human mesenchymal stem cells. That observation strongly suggests that FAZF modulates the regulation of osteoblastic differentiation via the BMP-2 pathway (Ikeda et al. 2007).

---

[16] Cancer endogenous to the bone marrow that affects the myeloid line of hematopoietic precursors cells, from which arise both osteoclasts and monocyte/macrophages, as well as red blood cells and platelets. Patients with this form of cancer evince a symptomatology that includes bone and joint pain, and increased risk of infection, as well as fatigue and other "sickness behaviors" associated with impaired innate immunity, and shortness of breath, easy bruising, and bleeding.

[17] Also denoted Zinc finger and BTB domain-containing protein-32, because it is a protein that in humans is encoded by the *ZBTB32* gene.

[18] CBFA1 is a key transcription factor associated with osteoblast differentiation and skeletal morphogenesis. As a member of the RUNX family of transcription factors, it has a Runt DNA-binding domain. It contributes to the scaffold for nucleic acids and regulatory factors involved in skeletal gene expression.

Chapter 3
Translational Evidence-Based Interventions in Osteoimmunology

3.1 The State of the Art

3.1.1 Research Synthesis

We began this monograph by citing an Evidence Report Publication (No. 07-E012) of the Agency for Healthcare Research and Quality (AHRQ). AHRQ plays a leadership role in the evolution of the field, from fundamental methodological issues to the design, establishment, verification, and dissemination of Registries. AHRQ defines Registries as: "... (a) patient registry is an organized system that uses observational study methods to collect uniform data (clinical and other) to evaluate specified outcomes for a population defined by a particular disease, condition, or exposure, and that serves one or more predetermined scientific, clinical, or policy purposes... registries (are) created for one or more of the following purposes: to describe the natural history of disease, to determine clinical effectiveness or costeffectiveness of health care products and services, to measure or monitor safety and harm, and/or to measure quality of care....(and) are classified according to how their (patient) populations are defined..." (Gliklich and Dreyer 2010).

As important as defining *what* is meant by a registry in the context of evidence-based and comparative effectiveness, decisions are the criteria by which we might evaluate the reliance of any given registry for that purpose. Here again, AHRQ sets the tone, as it were, and states in no uncertain terms: "...although registries can provide useful information, there are levels of rigor that enhance validity and make the information from some registries more useful for guiding decisions than the information from others. The term "quality" can be applied to registries to describe the confidence that the design, conduct, and analysis of the registry can be shown to protect against bias and errors in inference – that is, erroneous conclusions drawn from a registry...." (Gliklich and Dreyer 2010).

F. Chiappelli, *Osteoimmunopathology: Evidence-Based Perspectives from Molecular Biology to Systems Biology*, DOI 10.1007/978-1-4419-9495-0_3,

In evaluating the quality of registries for evidence-based and comparative effectiveness analyses, it may behoove the decision-maker to consider that the quality of a registry may be quantified from either of two fundamentally different vantage points:

- *Basic elements of good practice*: for example, a checklist that should be considered for all patient registries.
- *Potential enhancements to good practice*: viewed as the potential to strengthen the yielded information value in particular circumstances.

As the Gliklich and Dreyer (2010) report concludes: the validation, utility and use of a registry will depend largely upon the clinical problem and the patient population it serves (i.e., disease area), the type[1] and purpose[2] of the registry, and feasibility (i.e., concern of efficacy) and affordability (i.e., concern for cost-effectiveness). It is necessary, at this juncture, that we clarify the process and use of such registries and evidence reports.

First and foremost, it must be noted that the incorporation of "evidence reports" in the context of osteoimmunology and osteoimmunopathology reflects the integration of two sciences: the science of osteoimmune interactions and pathologies, as we have discussed up to this point, and the science of research synthesis for obtaining the *best available evidence*. Evidence reports, which are also referred to as evidence reviews in some circles,[3] state and summarize the process of research synthesis that yielded the best available evidence for treatment of a given condition (Chiappelli 2010) (Fig. 3.1).

Research synthesis follows the scientific process (research question/hypothesis – design and methodology – data analysis – inference), and can be outlined as follows:

- *Statement of the research question*: the question is crafted based on descriptors of:
  - The clinical problem and patient population (P)
  - The clinical interventions (I) under
  - Consideration/comparison/contrast (C)
  - Clinical outcome (O) of interest: PICO. The PICO question may undergo minor changes and alterations, as per the specific research question, it may examine a
  - Predictive (P), rather than a comparative model (hence, PIPO); or it may incorporate
  - Elements of time (T)
  - Settings (S) (hence, PICOTS, PIPOTS)

  In all instances, the research synthesis process starts by the statement of the research synthesis question along the cardinal domains just mentioned. The research question, as is the case in any scientific process of inquiry, when

[1] "...for example, product registries include patients who have been exposed to biopharmaceutical products or medical devices. Health services registries consist of patients who have had a common procedure, clinical encounter, or hospitalization. Disease or condition registries are defined by patients having the same diagnosis, such as cystic fibrosis or heart failure..." (Gliklich and Dreyer 2010).

[2] For example, the purpose of registries may include "...internal, external, or historical comparison groups in order to strengthen the understanding of whether the observed effects are indeed real and in fact different from what would have occurred under other circumstances..." (Gliklich and Dreyer 2010).

[3] AHRQ, for example, refers to *evidence reports*; whereas the American Dental Association, for instance, uses the term *evidence reviews*.

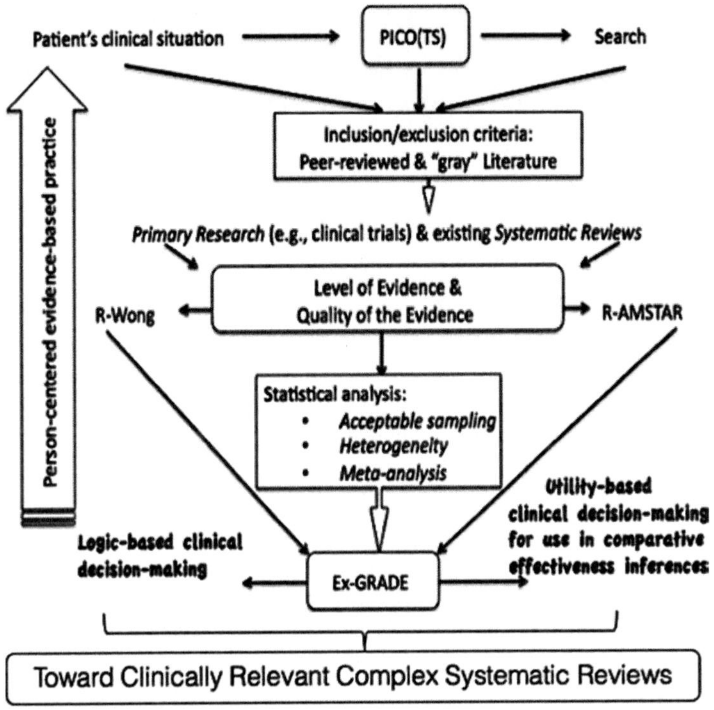

Fig. 3.1 The research synthesis process. The figure details the research synthesis as it begins from the patient's clinical situation. The clinical case is translated into a specific research question that specifically describes the patient's condition and history (P), the clinical interventions (I) under consideration (C), the clinical outcomes (O) under consideration in a given contextual framework of time (T), and clinical (or other) settings (S) (i.e., PICOTS). The figure also indicates how the PICOTS elements drive the search of the available peer-reviewed and "gray" literature, and the extent to which this process, which is aided by multiple search engines, is refined and made specific by means of inclusion/exclusion criteria that serve to define precise medical subject heading (MeSH) keywords. As the process continues, the available evidence that specifically responds to the PICOTS study question is examined for the level and the quality of the evidence by established and validated instruments, such as, the revised Wong scale (R-Wong) for fundamental studies, and the revised AMSTAR (R-AMSTAR) for systematic reviews. The quantification of the quality of the evidence permits acceptable sampling analysis and the extraction of the data from the individual reports permits, following heterogeneity/homogeneity testing, meta-analysis with fixed effect, or random effects inferences. Quality of the evidence data also converge into the analysis based on the expanded GRADE (Ex-GRADE), which congeals quantification of the quality of the evidence and of the strength of the clinical recommendations. Thus, is obtained both a qualitative and a quantitative consensus statement of the best available evidence specifically directed to respond to PICOTS. The figure also indicates how this best available evidence is then used in a logic-based clinical decision-making framework to provide the patient with optimally personalized efficacious treatment intervention in an evidence-based practice context. Alternatively, the best available evidence can be utilized in a utility-driven decision-making process to ascertain the most effective treatment intervention from the perspective of cost, benefit, and risk ratios. The design proposed in the figure lays the foundation for systematic reviews that can build on each other as the scientific literature continues to expand, while always retaining stringent checks and balances criteria. By this approach, systematic reviews of systematic reviews will be elaborated that will retain the three essential criteria of: (a) perusing all of the available evidence, (b) retaining the best evidence, in quantifiable and statistically analyzable terms, and (c) hinging on clinical relevance. The figures argues that these clinically relevant systematic reviews of systematic reviews, called "clinically relevant complex systematic reviews," increasingly find a place of prominence in osteoimmunology

stated in the affirmative mode – rather than in the inquisitive mode – becomes the hypothesis. That is to say, as in any other investigational pursuit that follows the scientific process, research synthesis is hypothesis-driven.

- *Methodological issues* are critical in research synthesis, as they are in any other research endeavor in science, in order to obtain the *best available* evidence. Methodological issues include: the sampling process, the measurement process, as well as selection bias and systematic errors.
  - The *sample* under study in research synthesis is not made of human subjects, animals, or cells in culture. Rather, the sample of a piece of research synthesis is the research literature itself. The research literature can be published or not published. The latter is often excluded, in part because it is exceedingly difficult to obtain in a valid and reliable manner, and in part because it has not been sifted through the widely accepted peer-review process. The former consists of two primary domains: the literature published following peer-review, and the literature available through the proceedings of scientific meetings, dissertations, and nonpeer-reviewed journals. The latter is termed "gray literature," and is most often excluded from research synthesis endeavors. That is to say, the research synthesis is most often focused, otherwise indicated, on the peer-reviewed literature. The search for that sample is obtained by utilizing the medical subject headings (MeSH terms) and keywords that can be derived from the PICO question – hence, the quality of the PICO question determines the quality of the sample. The search is actualized by accessing the National Library of Medicine (Pubmed-Medline, http://www.ncbi.nlm.nih.gov/pubmed), and usually at least two other search engines (e.g., Cochrane,[4] http://www.cochrane.org; Bandolier; http://www.jr2.ox.ac.uk/bandolier; EMBASE, http://www.embase.com; Center for Review Dissemination; http://www.york.ac.uk/inst/crd; google scholar; etc.). The purpose of the multiple search is to ensure comprehensive inclusion of all of the available literature, the bibliome, within the confines of the inclusion/exclusion criteria dictated by the research synthesis process, while at the same time minimizing as much as possible dangers of *selection bias*[5] and *systematic sampling errors*.

[4]The interested reader is advised to get on the mailing list of the Cochrane journal club (http://www.cochranejournalclub.com).

[5]Some degree of selection bias is unavoidable because of the very nature of our peer-review system. For example, a certain degree of publication bias cannot be avoided simply because, as a general rule, papers that are statistically significant, whether they demonstrate clinical relevance or not, tend to be prefentially published in the scientific literature, compared to reports that demonstrate clinical relevance but fail to reach statistical significance. The problem of publication bias is inherent to our present system of scientific literature, and is an unavoidable issue of the research synthesis process. The effect of the preferential acceptance of articles reporting significant results on research is critical: bias in favor of studies showing significant results alter the reliability of systematic reviews by reducing the included number of papers with opposing results. Since the validity of this type of publications depends on the representativeness and soundness of the source material, underrepresented evidence will have a disproportionally decreased influence on the outcome. That will be particularly grave when, as is the case for instance in osteoimmunology, research synthesis is utilized to obtain the best available evidence for treatment of pathologies in order to perform either evidence-based clinical decisions, or comparative effectiveness analysis (cf., Chiappelli et al. 2010a, b).

- Another critical aspect of the methodology of research synthesis pertains to the *assessment of the best evidence*, once, through the sampling process, we have obtained all of the *available* evidence. Each identified report must be evaluated for the level of evidence and the quality of the evidence (*vide infra*). The reliability of these assessments is ensured by the fact that they are obtained by means of well-crafted and validated instruments, and through two or more independent readers, whose replicability is tested and verified statistically (e.g., inter-rater reliability, coefficient of agreement[6]).

- *Analysis of the combined data* is a critical step in scientific research in general, and in research synthesis in particular. Whereas the level of evidence is usually a qualitative statement, or at best a semiquantitative ranking (e.g., level of evidence II-a, based on certain criteria; *vide infra*), the quality of the evidence generally produces some numerical values based on established scoring modalities. The quality of the evidence is best obtained by means of fully validated instruments for that purpose, although, in certain cases, in-house instruments can be used as well. The quality of evidence scores of all of the identified papers for a given PICO question can be combined into an *acceptable sampling analysis*, which is designed to identify and retain the highest scoring (i.e., the *best*) literature, and to reject the lowest scoring reports. Typically, low scores characterize reports with egregious deficiencies in research design, methodology, and statistical analysis of the data; and it is not arduous to grasp why the conclusions generated by these deficient resports ought not to be applied to the treatment of patients. Data can then be "extracted" from each of the reports within the pool of identified *best* literature, so long as the data address the same identical facet of the clinical outcome (O) under study (West et al. 2010; Ajaj et al. 2011). When that is the case, the studies are said to be *homogeneous*, a property that can be estimated statistically. The size of the differences between the experimental and the control groups (i.e., effect sizes) can be thusly extracted from each of the reports, and similarly the proportions of relative risk, and risk ratios, and tabulated and analyzed statistically together. When that is done – provided that the assumption of homogeneity is verified – then the overall, *overarching analysis* crossing over all the studies – the meta-analysis – will benefit from much increased sample size, compared to any individual study in the analysis, and thus proffer greater statistical power (i.e., detecting a statistical effects, if there one to be found). In brief, the data analysis of a research synthesis study is complex, and must ideally entail at least three steps:
  - Acceptable sampling analysis
  - Heterogeneity/homogeneity analysis
  - Meta-analysis

[6] It is important to note that inter-rater reliability obtains a correlation coefficient between two raters, and a high correlation implies that the two raters "agreed" on which item to score high or low (i.e., strong positive Pearson correlation coefficient). By contrast, Cohen's kappa coefficient is a statistical measure of agreement, which assesses whether or not the probability of the raters agreeing is larger than chance alone. That is to say, the Pearson intra-rater reliability coefficient is distinct from Cohen's kappa coefficient, although both values establish the degree of agreement between two raters, they are two distinct sides of the same coin.

It should be obvious, for instance, that omission of the preliminary acceptable sampling analysis will result in the potential inclusion in the meta-analysis of good as well as of sub-par research reports, which will undoubtedly dilute the statistical power of the meta-analytical step by incorporating extraneous systematic error (i.e., variability, variance). Similarly, if the homogeneity analysis is omitted, a meta-analysis will result that compares apples to oranges, yielding, again for the same reason, reduced power (Bartolucci and Hillegass 2010; Ajaj et al. 2011).

- That is to say, therefore, that the process of obtaining and utilizing the *consensus of the best available evidence* can only be attained if the steps outlined above are stringently followed:

  - Well-crafted *PICO question*, yielding appropriate keywords
  - Comprehensive *search* of the *available* literature bibliome
  - Accurate evaluation of the *level of the evidence*, and reliable quantification of the *quality of the evidence*
  - Accurate *acceptable sampling analysis*, *homogeneity analysis*, and *meta-analysis* of the data
  - *Consensus* of the reported in the *best available* literature
  - *Utility-based decision-making* for comparative effectiveness and *logic-based decision-making* for evidence-based intervention
  - Dissemination and updates to ensure increased *health literacy* and care

Only in this manner, will a definitive consensus of the statistical strength of the observed outcome be obtained, interpreted, and presented in a concerted and organized manner by means of a published report, which is referred to as a systematic review. Interpretation and utilization of the best available evidence presented in a systematic review can address two fundamentally diverse domains.

- When the best available evidence is utilized in a *logic model of decision-making*, it is usually for the purpose of conjoining it with the medical and dental history of the individual patient under care, his/her needs based on the clinical diagnosis and expertise of the treating physician or dentist, insurance coverage limitations and/or private financial ability/inability to pay for the recommended treatment, and expected side-effects. When the best available evidence is incorporated in such a manner in judicious clinical decision-making based on a logic inductive/deductive optimal treatment personalized for a given patient, that is aimed at determining what works best for a given patient – i.e., the efficacy of treatment. It is the core and substance for *evidence-based decision-making* (Chiappelli and Cajulis 2009; Chiappelli et al. 2009).
- When, on the other hand, the best available evidence is utilized in a markovian-type decision-making tree, where what is computed are the odds of a beneficial outcomes, often examined vis à vis cost (i.e., cost-benefit ratio; risk-benefit ratio), then the process of decision-making becomes evidently probability- and utility-based, rather than personalized patient-based. The *utility*

model of clinical decision-making has great value, particularly in the context of testing the effectiveness of treatment – hence, it is the ideal model of *comparative effectiveness*[7] *analysis* (Baio and Dawid 2008; Chiappelli and Cajulis 2009; Chiappelli et al. 2009).
- Other elements of clinical decision-making can come into play, which are discussed in an up-coming section (*vide infra*: Sect. 3.1.3).

In both instances, the decision-making process is aided by a carefully crafted summary statement, an executive summary of sorts, of the research synthesis process. These summary statements are necessary because, most often, the decision-maker may not have the time or expertise to read through and to evaluate each operative steps of the research synthesis as just described. They might simply want to be able to verify that the essentials have been covered adequately, and that, in finis, the research synthesis yielded this or that consensus of the best available evidence. These summary statements, called evidence reviews or evidence reports, are generally limited to 750–1,000 words, such as to permit quick reference by the decision-maker. It is important to state with emphasis that the generation of these summary statements of systematic reviews for the purpose of *dissemination and increased health literacy* is done by an expert in the field, fully trained in conducting systematic reviews, and in generating these summaries. Moreover, these summary statements are duly verified by a panel of experts in research synthesis and in the specific clinical field, prior to their dissemination to clinicians or decision-makers.[8]

In brief, the process of research synthesis as it integrates fundamental research pertains to the development and dissemination of methods for designing, conducting, analyzing, interpreting, reporting, and applying systematic research synthesis. The scope of the research synthesis methods extends to all aspects of the methods for conducting research synthesis, including literature retrieval and information science, data extraction, statistical methods, empirical research and simulations, software, graphics and reporting of synthesis results, issues of study quality, reporting or other systematic biases, narrative analysis and synthesis of qualitative data, synthesis of individual participant data, as well as the use of synthesis for developing practice guidelines and for integration with cost-effectiveness or decision analysis (Montori and Guyatt 2008; Chiappelli et al. 2010a; Ajaj et al. 2011).

---

[7] The Institute of Medicine Committee on Comparative Effectiveness Research Prioritization defined (2009) comparative effectiveness research and analysis as "the generation and synthesis of evidence that compares the benefits and harms of alternative methods to prevent, diagnose, treat, and monitor a clinical condition or to improve the delivery of care. The purpose of CER is to assist consumers, clinicians, purchasers, and policy makers to make informed decisions that will improve health care at both the individual and population levels."

[8] For example, the interested reader is referred to a recent AHRQ report: Creating a Framework for "Best Evidence" Approaches in Systematic Reviews, Review Protocol. September 2010. Agency for Healthcare Research and Quality, Rockville, MD. http://www.ahrq.gov/clinic/tp/bestevtp.htm.

In a practical sense, research synthesis emerges from the observations by a clinician, which is then translated into a PICO question, thus engendering the process just described. The process culminates in the consensus of the best available evidence, which, as noted, can be interpreted and incorporated either in an evidence-based or in a comparative effectiveness decision-making process. Of course, the clinician might also, and in fact very often does obtain biopsy samples or body fluids for further testing, in addition to the clinical diagnostic criteria (e.g., imaging CT, MRI, X-rays, etc.). When these biopsies and fluids are brought to the laboratory, they may be used for diagnostic research purposes, for example, characterization of the epigenetic proteomic signature – and thus yield novel information regarding the specifics of the pathology afflicting that patient. This new specific information can be incorporated, for example in the form of new key words in the search process, in the research synthesis, and thus may in the end contribute to yield a consensus statement that is better suited, more personalized to the specific patient case. In this manner, laboratory fundamental research contributes to enrich and to strengthen the research synthesis process in what is called translational[9] evidence-based decision-making (Chiappelli 2010; Ajaj et al. 2011).

What also has become evident over the past decade is that systematic reviews begin to accumulate in the literature, such that, for a given PICO question, several systematic reviews may be uncovered. That situation will become apparent in the subsequent section. A procedure is emerging, which was not utilized in this writing because it is still under development, which seeks to combine systematic reviews, such that a common and overarching consensus statement can be drafted. These combinatory systematic reviews are referred to as "complex systematic reviews" in the literature. However, and to distinguish complex systematic reviews that yield a clinically relevant consensus of the best available evidence, the term clinically relevant complex systematic reviews (CRCSRs) has also been coined (Chiappelli et al. 2010a, b). Later in this monograph, we discuss the concept of the strength of the clinical recommendation quantified by the Grading of Recommendations, Assessment, Development and Evaluation (GRADE) instrument, and its recent extension (Ex-GRADE), toward the further evolution of the field into translational CRCSRs (T-CRCSRs) (*vide infra*).

The protocol for CRCSRs also follows the research synthesis procedure, although it is a bit more complex: from the PICO question a search is generated that is limited[10] to including systematic reviews. The level of evidence of systematic reviews is, by definition, optimal; however, the quality of individual systematic reviews may vary.

---

[9] Fundamentally, and as defined by the National Institutes of Health (NIH), translational research is used *to translate* the findings in basic research efficiently into clinical practice.

[10] The clinically relevant complex *mixed* systematic reviews (CRCMSRs) combine (i.e., "mix") the traditional systematic review approach outlined above with the systematic reviews performed on a set of systematic reviews (i.e., CRCSRs). That mixing of two heterogeneous bodies of research in a single research synthesis process engenders significant analytical challenges and interpretative difficulties, which are still under development by my research group and others.

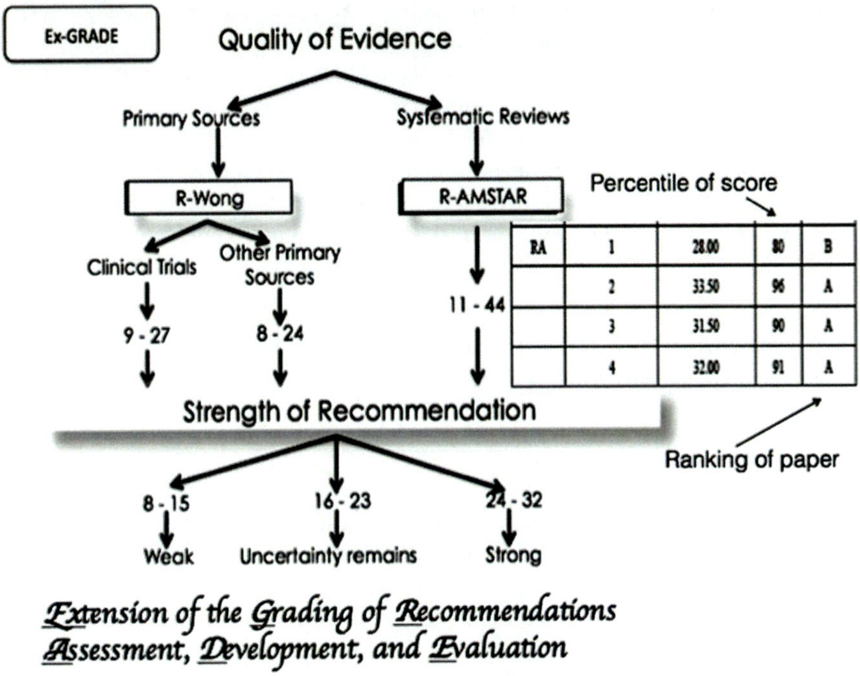

Fig. 3.2 Using the extension of GRADE (expanded GRADE, Ex-GRADE). The Grades of Recommendation Assessment, Development and Evaluation (GRADE) guidelines arose from a body of research synthesis work that has been established since the mid-1990s to early 2000s. By the mid-2000s, the GRADE had acquired its present structure consisting of two principal components: (a) grading the quality of the evidence, and (b) grading the strength of the clinical recommendations. The figure presents the structure of the extended or expanded GRADE, which integrates a fully quantifiable means to assess the quality of primary research or of systematic reviews, depending on whether the instrument is to be used to conduct a systematic review or a systematic review of systematic reviews. The Ex-GRADE also consists of a new quantifiable instrument for grading the strength of the recommendation, which has been fully validated (Phi et al. in press). The Ex-GRADE also integrates a system for ranking research reports in a clinician-friendly approach, such as that proposed by Kung et al. (2010). The insert of the figures shows a table that lists four systematic reviews on rheumatoid arthritis (RA) that were evaluated with the R-AMSTAR, whose total score is listed. Based on those score values, a percentile ranking is obtained, which is translatable to a letter grade. An approach similar to this is obtained with the overall Ex-GRADE scores (Barkhordarian et al. 2011)

Thus, it must be evaluated by means of validated instruments, such as the AMSTAR, which we have recently revised in order to yield quantifiable measures (cf., R-AMSTAR; Kung et al. 2010). Whereas R-AMSTAR scores permit acceptable sampling analysis and ranking of systematic reviews derived from the percentile of the individual quality scores (cf., Fig. 3.2), the problem remains unsolved as to how an overarching meta-analysis can be generated with sufficient statistical power and stringency across diverse systematic reviews. It is apparent that a simple cumulative meta-analysis, as has been proposed, may not satisfy the statistical requirements for quality research synthesis; rather, it appears that a Bayesian model of meta-analysis may be the sole and better approach to yield as reliable consensus of the best available evidence in CRCSR's (Oakley and O'Hagan 2004; Chiappelli 2010; Chiappelli et al. 2010a, b). Studies are ongoing in our research group at the moment, and it is premature to propose and de fend this or that protocol. Therefore, simplified evidence reports (i.e., evidence reviews) of individual systematic reviews are presented in the subsequent section, in order to illustrate the role of research synthesis in evidence-based and comparative effectiveness decision-making in oestoimmunology.

3.1.2 Example of Existing Systematic Reviews in Osteoimmunology

3.1.2.1 Rheumatoid Arthritis

Rheumatoid arthritis and other inflammatory diseases of bones and joints are, as noted in the previous chapter, particularly painful, cumbersome to treat, and exacting upon the aging and aged population. Often, patients self-treat, or engage in modalities that are considered alternative or complementary to traditional Western medicine. There are many over the counter alternative medicines available purporting to ease pain associated with arthritis, however there is no evidence supporting benefits for most alternative treatments including: vitamin A, C, and E, ginger, turmeric, omega-3 fatty acids, and chondroitin sulfate and these are thus not recommended. Glucosamine may have some benefit, and we have conducted a systematic review that demonstrated that S-Adenosyl methionine can relieve pain similar to nonsteroidal antiinflammatory drugs (Hardy et al. 2003).

Moreover, according to the philosophy of traditional acupuncture, energy circulates in "meridians" located throughout the body. Pain or ill health happens when something occurs to cause this meridian energy circulation to be blocked. The way to restore health is to stimulate the appropriate combination of acupuncture points in the body by inserting very thin needles. Sometimes in painful conditions, electrical stimulation along with the needles is also used. According to acupuncture theory, one way you can tell that acupuncture is relieving pain is that, you may feel numbness or tingling, called the qi, where the needle is inserted. However, acupuncture, as an adjuvant to an exercise-based physiotherapy program, did not emerge from a rigorous systematic review of the evidence in any greater improvements than the exercise program alone, or long-term benefits (Manheimer et al. 2010).

In a somewhat affine context, we might recall that hyaluronic acid, a naturally occurring anionic, nonsulfated glycosaminoglycan distributed widely throughout connective, epithelial, and neural tissues, is endowed of multiple beneficial functions, and has also been used to treat osteoarthritis. Hyaluronic acid-based "visco-supplementation treatment" interventions may have some merit when administered as a course of injections into the knee joint, perhaps by favoring the viscosity of the joint fluid – although the wide-spread functionalities of this compound suggests a more complex biological mechanism and set of side-effects. Simplistically however, one may suggest that visco-supplementation of hyaluronic acid may be efficacious in lubricating and cushioning the joint, thus minimizing friction and producing, overall, a decrease in inflammation, and consequentially a relief of pain. Whereas the hypothesis that hyaluronic acid might induce, in and of itself, an analgesic effect proper is questionable, immunological evidence supports a biochemical mechanism for the antiinflammatory function of hyaluronic acid.

Indeed, production of hyaluronic acid rises at sites of inflammation, often correlating with the accumulation of leukocytes (Day and de la Motte 2005). Moreover, platelets express the enzyme hyaluronidase 2, which cleaves hyaluronic acid into further fragments that are specific for inflammatory and angiogenic signaling (de la Motte et al. 2009), in fact confirming that hyaluronic acid, platelets, and monocytes constitute a newly identified triad of the inflammatory process (Menezes et al. 2009), with direct implications to osteoimmunology.

Mechanistically, biochemical evidence suggests that hyaluronic acid can be organized into a wide variety of molecular architectures by its association with specific binding proteins. This can lead to the formation of fibril-like structures that consist of several hyaluronan chains, which may cross-link thereupon promoting adhesion of leukocytes to these complexes and impeding leukocyte utilization of inflammation-promoting receptors. Therefore, leukocyte activation is blocked or blunted by these hyaluronic acid superstructures. These scaffolds may also prevent the loss of extracellular matrix components if the inflammatory process has commenced, and thus may contribute to the sequestering of proinflammatory cytokines and other mediators (Day and de la Motte 2005).

Hyaluronic acid binds to CD44 on lymphocytes, which is itself a marker of lymphocyte activation. Moreover, it binds to ICAM-1, and thus is likely to contribute to the control of ICAM-1-mediated immune cell adhesion and migration through the lymphatic system (Jackson 2009). Furthermore, and attesting to the role of hyaluronic acid as one of the conductors, or perhaps *the* conductor of the osteoimmune orchestra, is the observation that hyaluronic acid directly acts on osteoclasts by interfering with their differentiation in a Toll-like receptor 4 (TLR4)-dependent, but CD44-independent mechanism (Chang et al. 2007).

Clinical research on the beneficial outcomes of hyaluronic acid administration, particularly in the context of osteoarthritis remains inconclusive because its effectiveness and efficacy needs to be demonstrated. Therefore, a research synthesis protocol was designed to involve extensive search of the literature in Pubmed-MEDLINE, EMBASE, CINAHL, BIOSIS, and the Cochrane Controlled Trial Register. The outcome of interest was pain at rest, pain during or immediately after movement, joint function, and adverse side-effects. Twenty-two randomized controlled trials (RTC's) were identified, and synthesized in the systematic review. The assessment of pain by means of the visual analog scale is typically of weak reliability and validity, which led to measurements that were of usable quantitative value in many instances. Therefore, even though pain at rest seemed to be improved by hyaluronic acid, a clinically relevant observation, the data failed to yield a statistically significant outcome appropriate for further assessment. Similarly, no improvement in knee function was recorded quantitatively, but the effect of hyaluronic acid on knee function was clinically favorable. Adverse side-effects events occurred at a higher risk among patients who received the intervention (relative risk [RR] = 1.08, CI_{95}: 1.01–1.15). In brief, the best currently available evidence, while far from adequate, suggests that intra-articular hyaluronic acid has not been proven to be efficacious, or effective since it in fact may be associated with a greater risk of adverse outcomes (Arrich et al. 2005). Indeed, to this date, research synthesis of the best available evidence has not established the efficacy or the effectiveness of visco-supplementation of hyaluronic acid in the treatment of ankle osteoarthritis in the most current literature bibliome (Migliore et al. 2011), although in certain cases and for certain outcomes (e.g., function and patient global assessment at the 5–13 week postinjection period; Bellamy et al. 2005; or even at 14–26 weeks or sometimes longer postinjection period; Brzusek and Petron 2008) the efficacy of hyaluronic acid seems to be well established.

In a seminal AHRQ report on the treatment of primary and secondary osteoarthritis of the knee, Samson et al. (2007) compared the effectiveness of three modes of intervention: intra-articular visco-supplementation, oral glucosamine or chondroitin, alone or in combination, and arthroscopic lavage or debridement. The search, which did not exclude reports in languages other than English, and which included "gray literature" (i.e., conference proceedings), yielded data from 42 RCTs of visco-supplementation, all but one synthesized among six meta-analyses; 21 RCTs of glucosamine/chondroitin, 16 synthesized among 6 meta-analyses; and 23 reports on arthroscopy. Three domains of results were reported, which pointed to the act that the best available evidence failed to demonstrate clear and uncontroversial clinical benefit:

- *Visco-supplementation* trials generally showed positive effects on pain and function compared to placebo, although the overall clinical relevance could not be ascertained because of methodological weaknesses (e.g., variable trial quality, potential publication bias, questionable clinical significance[11]).
- *Glucosamine/Chondroitin* intervention, including careful assessment of the Glucosamine/Chondroitin Intervention Trial, a large (n = 1,583), National Institutes of Health-funded, multicenter RCT, showed no significant differences.
- *Arthroscopic lavage* with or without debridement emerged as equivalent to placebo.

3.1.2.2 Osteoporosis in Otherwise Normal Subjects

In the context of osteoporosis, most treatments have some proven efficacy in reducing the risk of vertebral fractures. Overall however, the evidence is less convincing in terms of the prevention of nonvertebral fractures, in part because

[11] These limitations emphasize the need for assessment of the level of the evidence and of the quality of the evidence, followed by an acceptable sampling step, as described above (*vide supra*).

available RTC's yield post-hoc subgroup analyses, rather than analyses based on the intent to treat. The intention-to-treat (ITT) analysis aims to circumvent the effects of crossover and drop-out, which alter the randomization to the treatment groups, and thus yield spurious data. In principle, the intention-to-treat analysis seeks to describe the potential effects of treatment policy rather than on the potential effects of a specific treatment, because it is a means of analyzing the outcome of a RCT based on the initial treatment intent, not on the treatment eventually administered. For the purposes of the intention-to-treat analysis, the entire sample that begins the treatment is considered to be part of the trial, whether they finish it or not (Lachin 2000).

A meta-analysis of eleven RCT's-Phase III was performed to compare the relative risks of nonvertebral antifracture efficacy for at least 3 years, confirmed by radiographs, of bisphosphonates, alendronate, and risedronate among several osteoporosis therapies in postmenopausal women. The research synthesis emphasized stringent assessment of the intention-to-treat sample. The analysis outcome established significant reductions in the relative risk of nonvertebral fracture for both alendronate (relative risk, RR = 0.86, CI_{95}: 0.76–0.97, $p = 0.012$) and risedronate (RR = 0.81, CI_{95}: 0.71–0.92, $p = 0.001$). Risedronate and strontium ranelate further evinced clinically relevant nonvertebral antifracture efficacy in the context of this research synthesis (Boonen et al. 2005).

To further test and establish the strength of the clinical recommendations and the cost-effectiveness of selective estrogen receptor modulators, bisphosphonates, and parathyroid hormone for the prevention and in the treatment of osteoporosis, a systematic review was brought forward with the specific goal of contributing to the body of knowledge aimed at preventing or reducing of osteoporotic fractures in postmenopausal women. The process of research synthesis emphasized meta-analysis, rather than the complete traditional systematic review structure. Studies were included in a random effect model meta-analysis, if the outcome of fracture incidence was reported in terms of the number of patients suffering fractures. The clinical decision-making model was derived from the meta-analytical results and inference to estimate the cost-effectiveness of osteoporosis interventions by calculating the number of fractures that occurred, as a function of the costs associated with the osteoporotic fractures, and the quality-adjusted life-years. As reference control, the conditions of breast cancer and coronary heart disease were modeled by the same approach because certain interventions can affect the risk probabilities for these conditions. Using a bibliome sample size of ninety RCTs that met the inclusion criteria, a comparison of five interventions (alendronate, etidronate, risedronate, raloxifene and teriparatide) to five reference control treatments (calcium, calcium plus vitamin D, calcitriol, hormone replacement therapy and exercise), and to a no-treatment placebo controls was possible. The intervention costs of treating all osteoporotic women for 5 years with alendronate, etidronate, risedronate, or raloxifene was elevated, and the cost adjusted per quality-adjusted life-years decreased dramatically with age. In fact, of the five tested interventions, only raloxifene appeared to reduce the risk of vertebral fracture in postmenopausal women, independently for low bone mineral density. However, the evidence indicated that none of the five interventions effectively reduces the risk of nonvertebral fracture in women, regardless of low bone mineral density, whereas they all led to substantial gains in quality-adjusted life-years, particularly among older women. The research synthesis also clearly evinced the estimated costs varied widely across the interventions, age, and clinical profile (i.e., prior fracture) (Stevenson et al. 2005).

Research synthesis, followed by a Markov model of probabilistic decision analytic techniques, was utilized to establish the comparative effectiveness and long-term costs and outcomes of five treatment and secondary prevention strategies for osteoporosis: placebo – "no intervention," alendronate, etidronate, risedronate, and raloxifene, in postmenopausal (65+ year old) osteoporotic women without prior fracture. Probabilistic sensitivity analysis, which is directed at predicting how changes in a given model inputs can, probabilistically, influence the outputs with the purpose of determining "good practice" (Andronis et al. 2009), was used to incorporate the impact of parameter uncertainty. In addition, deterministic sensitivity analysis, which differs from the former in that it seeks to predict how changes in a given model inputs can be determined by certain specified variables to influence the outputs, was used to compare alternative patient populations and modeling assumptions. Life years and Quality-Adjusted Life Years (QALYs, cf., notes) were the outcomes of interest in the comparative effectiveness investigation, which established that risedronate was less effective than etidronate and alendronate. Alendronate and etidronate were cost-effective alternatives for treating women with osteoporosis, although the model did not permit long-term projection (Goeree et al. 2006).

A related research synthesis confirmed these findings, as it examined the cost-effectiveness of nonfracture side-effects of osteoporosis treatments in women screened for osteoporosis at the age of 65, and treated osteoporotic subjects, as recommended with hormone replacement therapy, raloxifene, or alendronate. This approach utilized the Markov model of osteoporosis disease progression to simulate costs and outcomes by means of calculations of incremental cost-effectiveness ratios of screen-and-treat strategies, relative to a no-screen and no-treat strategy. Disease progression parameters, as well as cost and quality-of-life parameters were outcomes of interest in this analysis. Results showed that screening and treatment with hormone replacement therapy act in concert to increase costs and to lower quality-adjusted life-years (QALYs), relative to the no-screen, no-treat strategy, except when the model includes the assumption of no fracture, and thus no drug-related health effects. In conclusion, whereas raloxifene further increases costs and QALYs, alendronate emerges from this research synthesis as the most cost-effective strategy relative to the no-screen, no-treat strategy; in fact, with a fairly good and acceptable cost-effectiveness ratio (Mobley et al. 2006).

Fragility fractures cause significant morbidity and mortality. Effective osteoporosis treatment can reduce fracture incidence, but it is not known whether it is efficacious as well in reducing mortality. A research synthesis protocol was designed with the aim to determine whether effective osteoporosis treatment might be efficacious in reducing mortality. Two databases (i.e., Pubmed-MEDLINE, Cochrane Central Register of Trials), as well as some "gray" literature obtained from American Society for Bone and Mineral Research conference abstracts were consulted.

Eligible RCT's were included if they established efficacy interventions other than estrogen and selective estrogen receptor modulators in preventing both vertebral and nonvertebral fractures over a study duration longer than 12 months. Studies ($n=8$) of risedronate, strontium ranelate, zoledronic acid, and denosumab were included in the research synthesis. The consensus statement arising from the analysis was that treatment is efficacious in leading to a reduction in mortality (relative risk analysis I=0.89, CI_{95}: 0.80–0.99, $p=0.036$; relative risk analysis II=0.90; CI_{95}: 0.81–1.0, $p=0.044$), and that mortality reduction is independent from age or incidence of hip or other nonvertebral fracture. Treatment overall was actually all the more efficacious in reducing mortality among the older, frailer individuals with osteoporosis at high risk of fracture (Bolland et al. 2010).

In a study aimed to review the pharmacology, pharmacokinetics, pharmacodynamics, safety, efficacy, and use of denosumab in osteoporosis, breast cancer, prostate cancer, and multiple myeloma, pertinent research papers and abstracts were identified through a complete search of the two specific databases, Pubmed-MEDLINE and International Pharmaceutical Abstracts, for the publications between 1966 and July 2009. Key search terms included denosumab, its former name AMG-162 and its trade name Prolia, as well as its recognized functional characteristics as a fully human monoclonal antibody that specifically targets of the receptor activator of the nuclear factor-κB ligand (RANKL) system. Indeed, the FDA has given priority review status to the RANKL inhibitor denosumab (Prolia) to reduce skeletal-related events in cancer. For this particular systematic review, all available human clinical studies were included, except for studies in rheumatoid arthritis and giant cell tumor of the bone. The research synthesis design evinced that in patients with osteoporosis, denosumab significantly reduced bone resorption and consequential fractures, in large part by increasing bone mineral density, and reducing bone turnover markers of osteoclast-mediated function. Comparative effectiveness analysis determined that denosumab was at least as effective in reducing bone turnover markers as intravenous bisphosphonates in oncology patients. Efficacy analysis also established that patients with osteoporosis commonly reported side-effects of denosumab as arthralgia, nasopharyngitis, back pain, and headache. By contrast, the most common adverse effects of denosumab intervention in patients with cancer were infection often severe enough to require hospitalization, arthralgia, bone pain, and fatigue (Burkiewicz et al. 2009).

3.1.2.3 Osteoporosis in HIV/AIDS

Taken together, this research is particularly timely and critical to the HIV/AIDS pandemics. As noted above, HIV-seropositive adult men and women suffer from severely decreased bone mineral density, and increased risk of osteoporosis-related fragility fractures. Data indicate that the prevalence of osteoporosis in HIV-infected individuals is more than 3 times greater compared with HIV-seronegative control subjects. Moreover, HIV+ patients treated with antiretroviral therapy (ART) or protease inhibitors (PI) evince an even higher prevalence of reduced bone mineral density and increased risk of osteoporosis, compared with their respective controls. This evidence was reviewed in a research synthesis design involving a random effect meta-analysis. The search of the available evidence was broad and included the MEDLINE, PubMed, and EMBASE databases for peer-reviewed cross-sectional studies between January 1966 and November 2005. The PICO criteria included: pooled odds ratios of reduced bone mineral density and increased osteoporosis, as outcomes; the following patient population groups: HIV-positive vs. HIV-negative; and the following comparative interventions: ART-treatment vs. ART-naïve, and PI-treatment vs. PI-untreated. The twenty studies that met all of the inclusion/exclusion criteria yielded 884 HIV-seropositive patients, with a prevalence of decreased bone density of 67% (pooled odds ratio [OR]: 6.4), of whom 15% manifested clinical signs of osteoporosis (pooled OR: 3.7), compared to control HIV-seronegative subjects ($n=654$). Whereas studies did generally not correct for HIV/AIDS severity and treatment dose, regimen and duration, the data revealed that, compared with ART-naive patients ($n=202$, 10 studies), ART-treated HIV-seropositive patients ($n=824$) showed a 2.5-fold increased odds of decreased bone mineral density, and increased risk for osteoporosis (7 studies). Similarly, PI-treatment increased overall the risk for lower bone mineral density and greater risk of osteoporosis in HIV-seropositive patients (Brown and Qaqish 2006).

In a Cochrane systematic review, the effects of interventions aimed at increasing bone mineral density in HIV-infected adults were examined. Following a remarkable extensive search of the available evidence that included MEDLINE, EMBASE, LILACS, The Cochrane Library, as well as "gray literature such as Meeting Abstracts," AIDSTRIALS, ACTIS, Current Controlled Trials, National Institutes of Health Clinical Trials Registry, and Center Watch, only randomized trials that compared pharmacological or nonpharmacological therapy with placebo, no treatment, or an alternative therapy, in seropositive men and women 18 years of age or older were included. The clinical outcome of interest was increasing bone mineral density. The extensive and focused nature of the systematic review yielded three RCT studies that reported examination of the role of alendronate in patients with HIV and osteopenia or osteoporosis. A meta-analysis was precluded because of excessive heterogeneity ($p<0.0001$), which was attributable to marked differences in the study populations and interventions. Nonetheless, the sensitivity analysis showed that in two homogeneous studies (heterogeneity, $p=0.11$), alendronate, calcium, and vitamin D markedly improved lumbar bone mineral density after one year, when compared with calcium and vitamin D alone (weighted mean difference: +2.65; $CI^{95}=0.80$–4.51). Of note, is the clinically relevant observation that the alendronate-supplemented group did not evidence fewer fragility fractures (relative risk [RR]: 1.28; $CI^{95}=0.20$–8.21), or osteoporosis symptomatology (RR: 0.50; $CI^{95}=0.24$–1.01), and, overall, adverse occurrence of clinical outcomes was not significantly different between groups (RR: 1.28; $CI^{95}=0.20$–8.21). In one RTC, markedly heterogeneous with the others, demonstrated that patients with AIDS wasting showed clinically important improvement in lumbar bone mineral density following 3 months of testosterone enanthane treatment, compared to placebo, (weighted mean difference: +3.70; $CI^{95}=0.48$–6.92), whereas progressive resistance-training failed to improve this outcome in the patients. Neither the testosterone nor the resistance-training group suffered adverse effects. Taken together, the best available evidence that emerged from this systematic review confirmed that bisphosphonate therapy in HIV-seropositive adults, and testosterone in patients with AIDS wasting syndrome appears to be both safe and possibly effective to improve bone mineral density. Clearly, these conclusions are preliminary due to the limited number of pertinent and homogeneous studies included in this analysis (Lin and Rieder 2007).

In a related systematic review aimed at identifying the best available clinical data on bone ossification agents that may be considered for use in the treatment of osteoporosis and osteopenia in HIV-seropositive adults, an extensive search of the literature bibliome was performed of bisphosphonates, calcitonin, raloxifene, teriparatide, HAART, osteopenia, osteoporosis, and HIV/AIDS in human subjects. The search involved publications in English obtained through MEDLINE (1950-January 2008), EMBASE, as well as "gray literature" (i.e., abstracts from major HIV conferences; February 2001-October 2007). Outcomes of interest included pharmacology, pharmacokinetics, safety, and efficacy data for available treatment interventions with hormonal and nonhormonal agents. Three RCT clinical trials were obtained, and evaluated qualitatively for the use of a bisphosphonate in HIV-seropositive patients. A marked and clinically relevant increase in bone mineral density was observed in HIV+ patients taking alendronate, compared to the placebo groups, who received calcium, exercise, and/or vitamin D in one group or both. Conclusive consensus statements of the best available evidence could not be obtained, however, because of the restricted number of studies, and because of dosing restrictions that complicated the efficacious use of these agents; differences in diet, exercise, and calcium supplementation. Clearly, the question remains unanswered as to the clinical efficacy and effectiveness of alternate interventions, such as estrogen, testosterone, calcitonin, and teriparatide, which are less studied in HIV-seropositive patients.

3.1.2.4 Ankylosing Spondylitis

Fundamental research has suggested, as noted above, the relationship of TNF-α promoter polymorphisms with the clinical manifestation of ankylosing spondylitis. A meta-analysis collected the available bibliome of this topic, and focused specifically on the association between TNF-α promoter polymorphisms and disease susceptibility. Three search engines were consulted: Pubmed-MEDLINE, EMBASE, and Web of Science, and the numbers of individuals with various genotypes and alleles in both the case ($n=1,766$) and control groups ($n=2,114$) were extracted from the 14 identified relevant studies. CI_{95} of Odds ratios served to estimate the association. But, meta-analysis showed that there was no association between TNF-α polymorphisms and susceptibility of Ankylosing spondylitis in the overall population. In the HLA-B27+ subpopulation, the frequency of 308A allele decreased in patients with Ankylosing spondylitis (odds ratio$=0.721$; $CI_{95}=0.522$–0.995). But, the result was not statistically significant following exclusion of the Hardy–Weinberg equilibrium violation studies. We recall that the Hardy–Weinberg principle states that both allele and genotype frequencies in a population must remain constant and in equilibrium from generation to generation, unless specific artificial disturbing influences, such nonrandom mating, induced mutations, selection, limited population size, "overlapping generations," random genetic drift, gene flow, or meiotic drive, are introduced. In brief, therefore, whereas the convention that Ankylosing spondylitis is driven, or mediated by deregulation of the TNF-α pathway, the research synthesis protocol and meta-analysis of the data finds no relationship between the TNF-α promoter polymorphisms and ankylosing spondylitis susceptibility among patients with the HLA-b27+ make-up. The best available evidence shows no association between TNF-α promoter 238/308 polymorphisms and Ankylosing spondylitis susceptibility in both the overall population, and the HLA-B27+ population subset (Li et al. 2010).

Despite this fundamental evidence, it remains a clinically viable and important question to determine whether or not steroidal, nonsteroidal, or anti-TNF-α pharmaceutical agents are efficacious and effective treatment modalities for Ankylosing spondylitis. In one such study, the treatment effect of nonsteroidal antiinflammatory drugs for Ankylosing spondylitis was compared to TNF-α blockers (i.e., anti-TNF-α). The systematic literature research yielded RCT's that reported treatment efficacy in relieving pain and/or increasing physical function. Pooled effect sizes were estimated by meta-analysis, and interpreted by means of the fixed, and, when opportune, the random effect models. These two interpretative models differ greatly in that the fixed model holds the viewpoint that the outcomes (dependent variables) are fixed by, and nonrandomly dependent from the independent explanatory and predictor variables. By contrast, the random effect models posit that the dataset being analyzed consists of, and arises from a hierarchy of different populations, whose differences (i.e., variances) relate in a rather random fashion to that hierarchy. Indeed, when one assumes that given circumstances that drive and determine a given outcome (i.e., fixed model) can only and must arise from a random set of such possible circumstances, then it becomes evident that, in actuality, a fixed effects model is, simply stated, a special unique case among the random effects models. In the specific study at hand, the data were found to suggest that the treatment effects of NSAID's and anti-TNF-α were both clinically relevant, with a slightly greater efficacy of the anti-TNF-α intervention (Escalas et al. 2010).

Another research synthesis study was designed to assess the comparative clinical efficacy and cost-effectiveness of three anti-TNF-α agents, adalimumab, etanercept, and infliximab, for the treatment of Ankylosing spondylitis; electronic databases were searched, and gray literature was obtained as feasible in the form conference abstracts, economic evaluations reviews, and submissions to the National Institute for Health and Clinical Excellence (NICE, an independent UK organization that aims at providing evidence-based and comparative effectiveness guidance for promoting good health and preventing and treating ill health). Nine placebo controlled (i.e., conventional management of ankylosing spondylitis) RTC's were identified, which included two studies of adalimumab, five studies of etanercept, and two studies of infliximab. No study was identified that directly compared anti-TNF-α agents. Homogeneity of outcomes permitted meta-analyses on several means of clinical assessment for the condition, including general assessment of percent clinical improvement in Ankylosing spondylitis (ASAS), mean change in Bath Ankylosing Spondylitis Disease Activity Index (BASDAI), and mean change in Bath Ankylosing Spondylitis Functional Index (BASFI) at 12 weeks following initiation of anti-TNF-α therapy, compared to placebo, for all three drugs. Analyses were extended to 24 weeks for etanercept and infliximab. Meta-analyses of the 12-week data showed significant advantage of anti-TNF-α therapy over conventional placebo treatment: at 12 weeks, ASAS 50% responses were 3.6-fold more likely, and both BASDAI and BASFI scores were largely reduced, but the three pharmaceutical agents under examination were not dissimilar in efficacy. The cost-effectiveness conclusions were mixed, but suggestive overall of the trend that short time-frame anti-TNF-α therapies for Ankylosing spondylitis may not be cost-effective. The incremental cost-effectiveness ratios, that is the ratios of the change in costs of a given therapeutic intervention, compared to the placebo-alternative, to the change in effects of that intervention, were similar for etanercept and adalimumab, but lower for infliximab, particularly when corrected for quality-adjusted life year (QALYs, cf., notes). That is to say, the data of these meta-analyses revealed a substantial and quite unsustainable front-loading of costs of the anti-TNF-α interventions tested, which translates to indicate that none of the pharmaceutical agents considered in the study are cost-effective at current acceptable threshold, with infliximab yielding much poorer economic results overall. Certainly, the inference suffers from the fact that the short-term data available were used in a model intended to be predictive long-term and therefore speculative trends and parameter values were projected far beyond the available evidence. Sensitivity analyses are known to reveal wide variations in estimates of cost over the long term, despite the widely accepted dogma that it is unlikely that costs will decrease over time. Nevertheless, and in brief, whereas the best available evidence has established that three anti-TNF-α treatments under consideration are efficacious clinically in reducing the clinical symptomatology of Ankylosing spondylitis, this important study also unveils that the same interventions fail the test of comparative effectiveness. The short-term economic assessment clearly shows that none of the three anti-TNF-α agents is cost-effective at current acceptability thresholds. (McLeod et al. 2007). This systematic review analysis is a prime example of utilizing the same research synthesis protocol to uncover the best available evidence, which then demonstrates two disparate outcomes: satisfactory efficacy for evidence-based practice, and simultaneously unsatisfactory comparative effectiveness.

3.1.2.5 Osteonecrosis of the Jaw

In a systematic review designed to uncover the relationship between bisphosphonate use and development of osteonecrosis of the jaw in patients treated with oral bisphosphonates for the treatment of osteoporosis, an extensive search of the literature (i.e., MEDLINE, the Cochrane Database of Systematic Reviews, the Cochrane Central Register of Controlled Trials, EMBASE) yielded, following appropriate inclusion and exclusion criteria, 11 studies. Overall, the data evinced that while large numbers of patients with osteoporosis are treated with bisphosphonates, the relative prevalence of osteonecrosis of the mandible is relatively low, except for women aged over 60 years (Pazianas et al. 2007).

In another example, the treatment of ossifying fibroma was examined by systematic review to compare frequencies among four global groups. Two wide databases were searched, although perusal of three sources is generally recommended: the National Library of Medicine, PubMed-MEDLINE, and LILACS (search engine dedicated to Latin American and Caribbean Health Sciences; Biblioteca Regionàl de Medicina, BIREME). Inclusion of, for example WHOLIS, the World Health Organization Library Information System, might have been conducive to this systematic review. Inclusion criteria listed not only cases of ossifying fibroma that occurred in a series in the reporting authors' caseload, but also only those reports that confirmed fibro-osseous lesions histopathologically. The search yielded 64 reports (including the Hong Kong report) considered, of which 32 reports, for a total of 781 cases were included in the systematic review. Reports in languages other than English (ten) were not excluded, as were reports listing male and female patients despite the fact that women are generally affected more frequently than men. Patients with ossifying fibroma of the mandible were 3 times more prevalent than cohorts with this condition in any other anatomical site. Patients were generally young adults, with a mean age about 31, and a mode about age 40; but the mode female patient age was much in the preteen years. The major clinical pathology was recorded as being in 66% of all cases considered in the systematic review swelling and deformation of the jaws, and 84% of cases displayed buccolingual expansion, with, in potential downward displacement of the lower border of the mandible or, in close to all (90%) of maxillary cases, involvement of the maxillary antrum. Only 28% of reports included follow-up; and among these, a mere, but clinically significant 12% of cases recurred or showed signs of reactivated pathology, suggesting that long-term follow-up ought to be the norm for successful evidence-based treatment of ossifying fibroma (MacDonald-Jankowski 2009).

In a similar study by the same group, the principal features of focal cemento-osseous dysplasia were examined by means of a systematic review in order to compare frequencies among four global groups. The same two search engines were consulted, and as mentioned earier, only reports that described cases duly confirmed through radiographic and histopathological means were included. Ten reports constituted this systematic review. Fibro-osseous lesions predominantly affect females and the mandible. Two at-risk global communities emerged: those from East Asian, and those from Central Africa origin. Long-term follow-up of large series that would have revealed the long-term outcomes of focal cemento-osseous dysplasia is not available to date. This information is necessary for future well-informed evidence-based and comparative effectiveness treatment of edentulism in these regions of the world, and potential for osseointegrated implants (Macdonald-Jankowski 2008).

3.1.3 Clinical Decision-Making

People who exhibit value-induced bias- distorting relevant probabilities to justify medical decisions- may make suboptimal decisions, and may distort relevant probabilities to justify their preferred choices (Levy and Hershey 2008).

Decision analysis techniques can be applied in complex situations involving uncertainty and the consideration of multiple objectives. Models for decision-making require the estimation of a plethora of parameter and of their independent and interactive probabilities. The overarching purpose of these models is to enable applications and implications to optimizing either efficacy or effectiveness of healthcare interventions.

One experimental model has utilized a Java-based software resource, the Clinical Decision Modeling System, to implement Naïve Decision Modeling, specifically in the realm of performance evaluation measures to compare the cost-effectiveness of strategies for breast and lung cancer detection. This approach, even when assuming equal cost, emerged as a highly practical applied strategy to direct the process of establishing evidence-based integrative translational clinical research priorities; although the model evinced substantial limitations in the context of providing clinical decision support. In brief, the computerized model usefulness, at the present level of development, is limited to simplifying the objective-driven planning of complex integrative clinical studies without requiring a multiattribute utility function. It is probable that this informatics-based approach for clinical decision-making will evolve into an algorhytm that will permit efficient integrative translational clinical study designs that move beyond simple pair wise competitive studies for contrasting benefits of using alternative clinical combinations to affect strategic and cost-effective clinical workflows (Shi and Lyons-Weiler 2007).

Current models of medical decision making include shared decision making, informed decision making, and evidence-based choice. In these models, the choice of the most reasonable treatment option often requires acknowledging the potential for conflict between patients and their set of uninformed biases and preferences, and the clinicians' expertise and knowledge of the supportive evidence (Whitney et al. 2008).

In recent years shared decision-making has gained importance as an appropriate approach to patient–physician communication and decision-making. However, there is a conceptual variety that implies problems of inconsistent measurement, of defining relationships of shared decision making and outcome measures, and of comparisons across different studies. This article presents the results of a literature search of psychometric instruments measuring aspects of decision-making. Altogether 18 scales were found. The majority covers the patients' perspective and relates to preferences for information and participation, decisional conflict, self-efficacy as well as to the evaluation of decision-making process and outcomes. The scales differ widely in their extent of validation. Although this review is not exhaustive, it presents a variety of available decision-making instruments. Yet, many of them still need to show their psychometric quality for other settings in further studies (Simon et al. 2007).

Effective handling of uncertainty is one of the central problems in medical decision-making. The sources and effects of uncertainty in medical decision making can be quantified by means of branching probabilities (e.g., Markovian tree) and node utilities for probability schematas of alternative treatment strategies. Certain public domain software packages are now available, but are, at present, limited by the potentially unmanagable complexities of multavariate nature of the possible sequence of diagnostic activities, pathophysiologic variables of each individual patients, and the individualized, patient-centered evidence-based course of treatment (Dittus et al. 1989).

The construction and evaluation of *Markov decision processes* have traditionally been powerful analytical tools for sequential healthcare decision making under uncertainty Markov decision processes generalize standard Markov models in that a decision tree is embedded in the model, which permits multiple decisions over time. Furthermore, the Markov decision tree is superior to standard decision analysis in that it permits a faster and more reliable computation time for solving medical decisions (Alagoz et al. 2010).

The *transtheoretical model* of decision-making calls for stages of behavior change that can represent temporal dimension for behavior change. Therefore, the decision-making variables of the pros and cons of changing are foundational and systematic for the evolutions of behavioral relationships, and determinant for decisions to decrease health risk behaviors and increase health-enhancing behaviors (Prochaska 2008).

The core idea of *fuzzy trace theory* is that people rely on the gist of information, its bottomline meaning, as opposed to verbatim details in judgment and decision-making. In fuzzy trace decision-making, precise information about benefits of risks is not necessarily effective in encouraging prevention behaviors or in supporting medical decision-making. People can get the facts right, and still not derive the proper meaning, which is key to informed decision making. In addition, retrieval of health-related values and processing interference brought on by thinking about nested or overlapping probabilities of occurrence are also important (Reyna 2008).

The evidence-based practice process requires integrating the evidence with consideration of practical resources and patient preferences in collaborative and integrative engagement for making health-related decisions. This transaction is relevant both in the principal theories of clinician decision making (e.g., expected utility and fuzzy trace) and in the principal theories of patient health decision making (e.g., transtheoretical model and reasoned action for improved lifestyle), because these theoretical paradigms rest on similar data strands, that consist of evidence, resources, and preferences. The preponderance of these data supports computational approaches and the development of specific algorhythms, which still remains needed (Spring 2008).

In summary, ignoring evidence-based theory is no longer defensible in medical decision-making and health, regardless of whether the focus is on research or on practical applications. The three theories (theory of reasoned action, transtheoretical model, and fuzzy trace theory) is supported by empirical evidence, they present disparate views of risky decision-making, behavioral change, health promotion, and medical decision-making. The following points deserve reiteration:

- In all three theories, important aspects of decision making *need not be conscious*. Attitudes and norms need not be consciously deliberated to influence intentions, and, subsequently, behavior. People in the precontemplative stage are unaware that they are "underestimating the pros of changing and overestimating the cons" relative to perceptions at other stages of change. Decisions are mentally represented (as verbatim and as gist representations), and research has shown that gist representations often operate unconsciously.
- Each theory has a relatively clear position on what is *prescriptively desirable*. The theory of reasoned action can explain behavior that is reasonable, that emerges from well-ordered intentions – The theory of reasoned action has much evidence to support it; meta-analyses show that behavioral intentions are significant predictors of behavior. In fuzzy trace theory, as in the other approaches, internal coherence and good outcomes (e.g., good health outcomes) are considered generally indicative of better decisions.
- They can be applied to *both patients and healthcare providers*.

3.2 Directions for the Future

3.2.1 Translational Evidence-Based Interventions: Toward a Novel Model

The optimal translational framework for evidence-based and comparative effectiveness research must redefine the objective of translation from that of institutionalizing effective interventions to that of improving population and individual patient health. This outcome can only be obtained by influencing the determinants of health and disease in their totality and complexity, from the epidemiological, to the psycho-emotional and social science perspectives, from the proteomic signature to the metabolomic and physiomic contexts, from individual patient data to meta-analyses, in order to recognize and to appreciate that many types and facets of research contribute to the overall profile of the best available evidence, and the shaping of evidence-based policy, practice, and future research. The pivotal role for research and evidence synthesis cannot be overlooked in the process of advancing the field of applied translational evidence-based and comparative effectiveness decision-making (Ogilvie et al. 2009).

That is the reason why the translation and diffusion of findings into health care validate the potential of evidence-based innovation to improve clinical practice and affirm the benefits of society's investment in advancing science. Therefore, novel model of translational evidence-based interventions must now arise in health care in general and osteoimmunology in particular.

The translation of evidence-based approaches to communities and populations can de facto be actualized. Partnerships among diverse people and organizations can be obtained using a mix of approaches that work together to promote the transference of evidence from research into practice through local, regional, and national partnerships (Breslau et al. 2010). That collaborative and cooperative cross-feeding empowers the patients to become active participants in the evidence-based treatment, and challenges a dialog between fundamental researchers and clinicians to foster translational and trans-national modes of intervention (Ajaj et al. 2011).

3.2.2 *Perfecting the Protocol for Getting the Best Available Evidence*

That is akin to stating that, in every domain of health care, including osteoimmunology, the next decade will witness the felicitous integration of basic science as well as patient input in a fully patient-centered translational evidence-based and comparative effectiveness clinical decision-making process in a global scale. We are early in this process, and need at this point to examine the elements we presently have and what needs to be improved.

While the Promoting Action on Research Implementation in Health Services (PARiHS) framework is widely promoted to implement evidence-based clinical practices, it remains to be demonstrated what pool of validated measurement instruments the network in actuality possesses to operationalize the bold constructs it proposes. One such tool is recognized under the acronym ORCA, which stands for the 77-items Organizational Readiness to Change Assessment instrument. ORCA is structured and validated along 19 subscales and three primary scales that correspond to the fundamental core elements and subelements of the PARiHS framework:

- *Strength and extent of evidence* for the clinical practice changes represented by the QI program, assessed with four subscales (Cronbach $\alpha=0.74$).
- *Quality of the organizational context* for the QI program, assessed with six subscales (Cronbach $\alpha=0.85$).
- *Capacity for internal facilitation* of the QI program, assessed with nine subscales (Cronbach $\alpha=0.95$).

However, ORCA generally fails with poor reliability among measures of evidence, and factor analysis results for measures of general resources and clinical champion role. Thus, the PARiHS framework, while encouraging, still suffers from poor validity and reliability in certain factors and domains (Helfrich et al. 2009).

The PARiHS structures is effective is explicating a process that catalyzes new knowledge adoption and use by individuals and systems to solve problems. Each conceptual perspective suggests that translation is not complete until the extent and impact of use is examined and understood. The AHRQ framework has proposed that measures in addition to PARiHS in order to include changes in patterns of care and changes in policies, procedures, or protocols. This perspective supports the evaluation of impact of evidence-based practice using process measures that integrate clinician knowledge, actual performance of the practice, and patient/clinician outcomes (Donaldson et al. 2004).

It remains certain that the PARiHS framework has proved to be, and continues to be, for the most part, a useful practical and conceptual heuristic for researchers and practitioners in framing translational evidence. Indeed, it has been proposed that PARiHS may best be utilized as a two-stage process: one that initiates with a preliminary diagnostic and evaluative assessment of the evidence within a PICO context (vide supra), and that then progresses through aggregation and analysis of pertinent data to determine the consensus of the most appropriate mode of intervention in a decision-making mode (cf., previous section) (Ajaj et al. 2011). PARiHS thus provides a structure and a framework by which diagnostic and evaluative information come together to shape, craft, and mold the decision-making process for a given intervention targeted to the specific situation and context, and to the participating stakeholders (Kitson et al. 2008). As such, the construct validity of PARiHS is imbedded in the very contextual, epistemological, and ontological definition of evidence-based and comparative effectiveness in decision-making. The next generation of work in this are will see the full implementation of PARiHS in the context of osteoimmunology and osteaoimmunopathology.

In a parallel effort to perfect tools and instruments for obtaining and utilizing in clinical practice the best available research evidence, the Appraisal of Guidelines Research and Evaluation-Europe (AGREE 2003) is a widely adopted globally around the world evaluation instrument that is useful for the quantitative assessment of quality of the development of clinical practice a guidelines. AGREE proceeds along six domains:

- Scope and purpose
- Stakeholder involvement
- Rigor of development
- Clarity and presentation
- Application
- Editorial independence

Scores from multiple (at least two) assessors collected, evaluated, and reconciled if need be as above. Assessment are semiquantitative/semiqualitative in nature and obtain as: “strongly recommended for use in practice,” “recommended for use with some modification or proviso,” “not recommended as suitable for use in practice,” or “unsure.” In an effort to improve upon obvious weaknesses, the 23-item AGREE-II instrument was recently disseminated, which, besides using a seven-point scoring system, also evaluated the usefulness of each of the original AGREE items toward enhancing content validity (Brouwers et al. 2010a, b).

Moreover, the Grading Recommendations Assessment, Development and Evaluation (GRADE) system (*vide supra*; Schünemann et al. 2008; Brozek et al. 2009) provides the opportunity to obtain evidence-based recommendations that are based on an evaluation of the level of evidence, the quality of the evidence, and the clinical relevance of the best available evidence (Terracciano et al. 2010). Therefore, the GRADE instrument finds for itself a central niche in research synthesis for

evidence-based and comparative effectiveness clinical decision-making in that it pertains to the identification of the best available evidence for reporting in systematic reviews, and for integration in revised clinical practice guidelines.
Specifically, GRADE allows four grades of quality of evidence:

- High
- Moderate
- Low
- Very low

GRADE also yields three levels of the strength of the clinical implications and recommendation:

- Strong
- Weak
- Conditional to an intervention – pro
- Conditional to the intervention – con

Overall, one strength of GRADE is that the semiquantitative GRADE estimate is meant to yield a level of confidence that desirable effects predominate over untoward ones with a certain intervention. Another strong advantage of GRADE is that it permits recommendations to be formulated and the method to be clarified and made more explicit and transparent (cf., Fig. 1.2). It takes into account clinician expertise and patient values when defining and grading the relevant outcomes, thereby avoiding any influence from literature precedents; while this process may be more time-consuming than other related approaches, it clearly can be considered to be a notable strength of this method. In fact, GRADE, when compared to another widely used instrument, the Strength of Recommendation Taxonomy (SORT), rated equally as important to the critical and timely need in this domain of research synthesis science to evaluate the quality of evidence and the strength of recommendations (Faggion 2010), considering the fast proliferating and the gargantuan heterogeneity of the quality and type of evidence in health care, from dentistry, to medicine, nursing, nutrition science, psychopathology, and complementary and alternative healthcare practice. SORT ensures guidelines to obtain unified and transparent development of the best available evidence for clinical decisions, as well as to identify potential conflicts of interest, and offer flexibility in various clinical situations (Lin and Slawson 2009). This urgency has been recognized by World Health Organization (WHO) and many other organizations around the world.

A recent study revealed that only ten of 14 WHO and five of seven World Bank recommendations were consistent with the direction of effect claims, as emerging from research synthesis protocols and systematic reviews. Therefore, it is critical and timely that WHO, the World Bank, donor agencies, national governments, as well as insurance carriers improve their use of (or at least, their reporting about their use of) research evidence. Decision-makers and clinicians should critically evaluate the quality and local applicability of recommendations from any source, including international organizations, prior to their implementation (Hoffman et al. 2009).

The urgency perceived at a global and international level to identify the best available evidence for transparent, valid, and reliable identification of the best available healthcare practice in the first as in the third World is exemplified by the emergence of such networks as the international forum for evidence-based decisions and comparative effectiveness research (ifebdcer.org), and EQUATOR, the enhancing the quality and transparency of health research collaborative international initiative dedicated to enhancing the quality and transparency of health research set up specifically to advance high quality reporting of health research studies, and to promote good reporting practices including the wider implementation of reporting guidelines (Simera et al. 2009, 2010).

Clearly, GRADE is an important step forward in evaluating not only the best available evidence, but also, and especially, the strength of clinical recommendations that pertain to that evidence. It is a unique instrument at this particular stage of the evolution of the science of research synthesis, and of its implication to evidence-based and comparative effectiveness clinical decision-making. It actually congeals in a very powerful and unique manner the assessment of the level and quality of the evidence, with an assessment of clinical relevance. In recent developments, we have up-dated and expanded the interpretative realm of the GRADE, and produced the Ex-GRADE, which we have discussed in greater details elsewhere (Chiappelli et al. 2010a, b; Barkhordarian et al. 2011) (*vide supra*, cf., Fig. 3.2).

The level of evidence is established on the basis of the type of study design that was used to generate the evidence under evaluation (Chiappelli 2008; Chiappelli et al. 2010a, b). Typically, a hierarchy is generated as follows (cf., U.S. Preventive Services Task Force):

- *Level I*: Evidence obtained from at least one properly designed RTC.
- *Level II-1*: Evidence obtained from well-designed controlled trials without randomization.
- *Level II-2*: Evidence obtained from well-designed cohort or case-control analytic studies, preferably from more than one center or research group.
- *Level II-3*: Evidence obtained from multiple time series with or without the intervention. Dramatic results in uncontrolled trials might also be regarded as this type of evidence.
- *Level III*: Opinions of respected authorities, based on clinical experience, descriptive studies, or reports of expert committees.

The UK National Health Service uses a similar system with categories labeled A, B, C, and D

- *Level A*: Consistent Randomized Controlled Clinical Trial, cohort study, with clinical decision rule validated in different populations.
- *Level B*: Consistent Retrospective Cohort, Exploratory Cohort, Ecological Study, Outcomes Research, case-control study; or extrapolations from level A studies.
- *Level C*: Case-series study or extrapolations from level B studies.
- *Level D*: Expert opinion without explicit critical appraisal, or based on physiology, bench research or first principles.

In more recent years, since the fast emergence of systematic reviews, it is generally accepted that systematic reviews have a Level of Evidence that is even higher than I or A – a level "super-I/A." The complication of course arises at present, when one considers that the science of research synthesis continues to evolve, such that multiple systematic reviews on a given clinical questions can now be pooled into what has been referred to as "complex systematic reviews." Following in the logic above, complex systematic reviews ought to be recognized at a level of evidence even higher than systematic reviews – presumably, "super super-I/A."

The quality of the evidence is certainly as distinct from the level of the evidence as well of badly played baroque or rock music. The level of evidence pertains to what design was utilized to generate the evidence, to same extent as the musical compositional canon permits a classification of genres into medieval, classical, modern etc. The level of the evidence responds to the question "what was done to obtain the evidence?"

By contrast, the quality of the evidence is akin to whether or not that renaissance motet composed by Gabrielli or this Mahler symphony are correctly executed and well played. The quality of the evidence refers to the question "how well was the evidence obtained?"

In the search for the quality of the highest evidence, the consolidated standard for randomized trials (CONSORT[12]) were introduced (Begg et al. 1996; Jüni and Egger 2009), and have been continuously revised, improved, and widely used in this research synthesis for traditional Western healthcare (Schulz et al. 2010), as well as for acupuncture and complementary and alternative medicine, in the form of the standards for reporting interventions in clinical trials of acupuncture (STRICTA) (MacPherson et al. 2010).

To a similar aim, but directed to observational designs in the health sciences in general, and in epidemiological studies in particular, the STROBE statement was articulated for the purpose of strengthening the reporting of observational studies in epidemiology (von Elm et al. 2007). Individual checklists are available to evaluate the strength of cohort studies, cross-sectional studies, and case-controlled studies. The STROBE, in fact, has such merit that it was crafted by the STROBE group, a multinational team of academicians from Europe (e.g., Switzerland, UK) and North America (e.g., Canada, US) (strobe-statement.org) is used internationally, and is available in several languages (i.e., English, Chinese, Japanese, Spanish, Portuguese, German, Italian, Greek). This instrument was then expanded to assess the strengths and weaknesses of genetic, genomic, and proteomic evidence for the eventual integration of this information into evidence-based healthcare (i.e., strengthening the reporting of genetic association studies, STREGA; strega-statement.org) (Little et al. 2009; Thelle 2009; von Elm et al. 2009).

With respect to evaluating the quality of systematic reviews, Shea and colleagues developed and characterized the assessment of multiple systematic reviews instrument (AMSTAR), through a process of factor and cluster analyses of previously existing

[12] The CONSORT statement has recently been revised as CONSORT-2010 (*vide infra*).

instruments for this purpose (e.g., Overview Quality Assessment Questionnaire, OQAQ; Sacks' checklist; quality assessment of studies of diagnostic accuracy included in systematic reviews, QUADAS) (Shea et al. 2007, 2009). This process resulted in the identification of 11 domains that are sine qua non's of an adequate systematic review, and which constitute the 11 items of the AMSTAR:

- "A priori" design provided
- Duplicate study selection and data extraction
- Comprehensive literature search
- Status of publication (i.e., gray literature) used as an inclusion criterion
- List of studies (included and excluded) provided
- Characteristics of the included studies provided
- Scientific quality of the included studies assessed and documented
- Scientific quality of the included studies used appropriately in formulating conclusions
- Methods used to combine the findings of studies
- Publication bias
- Conflict of interest

Of course, we recognize fundamental domains of the scientific process in the list above. The initial item pertains to the research question (PICO question of the systematic review under evaluation: is it evident?, that is, was this piece of research synthesis a question-driven, a hypothesis-driven endeavor, or merely an ephimerous "shot in the dark," a "fishing expedition," a groundless exercise. The following six items pertain to the integrity of the research synthesis design per se: they address the specific aspects of replicability of measurement (duplicate study selection and data extraction), sampling methodology (comprehensive literature search, status of publication (i.e., gray literature) used as an inclusion criterion, list of studies (included and excluded) provided), and measurement (level of the evidence: characteristics of the included studies provided; quality of the evidence: scientific quality of the included studies assessed and documented). Data analysis is addressed in the following two items (scientific quality of the included studies used appropriately in formulating conclusions: that is, acceptable sampling analysis; methods used to combine the findings of studies: that is, meta-analysis). The scientific method demands that the process engenders inferences that are free of bias, which is addressed by the last two items of the AMSTAR (publication and conflict of interest bias).

In a recent study we alluded to above (*vide supra*), we revised the AMSTAR instrument, detracting nothing from its content and construct validity, and utilizing the very criteria employed in the development of the original tool, with the aim of yielding an instrument that can quantify the quality of systematic reviews. We validated the revised AMSTAR (R-AMSTAR), and presented its implications and applications in evidence-based clinical decision-making in health care (Kung et al. 2010) (*vide supra*; cf., Figs. 3.1 and 3.2).

Increasingly, therefore, systematic reviews include – must include accurate analysis of the data produced by the research synthesis process. Whereas the inclusion of acceptable sampling analysis is still in its infancy, as the field of evidence-based and comparative effectiveness decision-making in health care becomes better defined, meta-analysis, a statistical methods with a long history starting with Pearson and revived by Glass in the 1970s, is becoming almost a de facto necessity in any respectable systematic review. Over a decade ago, it became apparent that standards must be established for the appropriate reporting of meta-analytical studies, especially when these pertained to the identification of the best available evidence for health care. The Quality of Reporting of Meta-analyses (QUOROM) statement (Moher et al. 1999) presents a checklist, and a flow diagram to outline the optimal flow of presentation of the abstract, introduction, methods, results, and discussion sections of a report of a meta-analysis. They were structured and organized into 21 headings and subheadings, which had the advantage of providing a set of guidelines for investigators, but were often arduous to understand and follow for the neophites.

In a recent development, QUOROM was revised and improved, and presented as the Preferred Reporting Items for Systematic Reviews and Meta-analyses (PRISMA) statement (Liberati et al. 2009; Moher et al. 2009). Whereas longer and more complex, PRISMA (prisma.org), which consists of a 27-item checklist and a four-phase flow diagram, is actually more user-friendly than QUOROM.

This superior cutting edge tool is has already found its way in bone-related research, as it was the primary assessment instrument of a systematic review designed to test radiographic marginal bone level changes and the survival of platform switched implants compared to the conventional matched abutment implants. The research synthesis protocol and the meta-analysis showed that platform switching may preserve the inter-implant bone height and soft tissue levels, and that the degree of marginal bone resorption appears to be inversely related to the extent of implant abutment mismatch (Atieh et al. 2010). Moreover, and in the osteoimmune context of this writing, it is important to report the recent utilization of PRISMA in the identification of the best available evidence for the comparative assessment of antirheumatic drug (DMARD) single- and combination therapy, glucocorticoid therapy, and biologic therapy for controlling inflammation-driven joint destruction in rheumatoid arthritis. The data of the systematic review of 70 trials evinced the clear outcome in that DMARD's, glucocorticoids, biologics, and combination treatments significantly reduced radiographic progression at one year with a relative effect of 50–80% (Graudal and Jürgens 2010).

In conclusion, it is noteworthy to reiterate that the gap between research findings and practice has been, and continues to be, a concern for the international community. A number of descriptive studies have elucidated barriers and facilitators of evidence-based practice in nursing. It is argued that it is now time to use findings from these studies to design and test interventions that explicitly target barriers to the use of evidence in practice rather than doing further research to describe generic barriers and facilitators to evidence-based practice. This article discusses research methods to advance our knowledge regarding the efficacy of translating research into practice (TRIP) interventions that promote and hasten adoption of evidence in practice. Following systematic review of published research in a research synthesis protocol, exemplars of translation studies are singled out and discussed individually to highlight the benefits and challenges. It is evident that the process, while laudable for its unbiased research synthesis process, also suffers from bias inherent to selection in conducting TRIP studies. Moreover, limitations of TRIP modalities with respect to sample attainment and allocation, unit of analysis assessment, intervention characteristics, outcome measurement, and sustainability are mitigated only in part by the myriad of TRIP initiatives, increased identification and use of evidence in practice, attempts at obtaining systematically the best available evidence for the effectiveness of these initiatives (Titler 2004). It could be argued that TRIP initiatives may be put into place and evaluated most effectively in evidence-based practice centers (EBPCs) and the centers for education and research on therapeutics (CERTs) of the kind established by the AHRQ.

It is now timely and critical to conjoin advances in basic knowledge obtained by fundamental investigations of biological mechanisms (i.e., molecular and cell biology, animal and systems physiology) to research synthesis paradigms in the pursuit of translation evidence-based research, toward what might come to be termed as translational TRIP (T-TRIP). In that vision, T-TRIP would be conceptualized as TRIP from the overarching view of the translational process from the patient in the clinic to the research laboratory and the research bench back to the patient, as defined by NIH. In short, the process of TRIP considers strictly the utilization of clinical research (e.g., clinical trials) into practice-in a mode akin to the practice-based research networks (PBRN's). By contrast, T-TRIP is conceptualized as the process that seeks the integration of translational research into the practice reality, and emphasizes the reliance on translational clinically relevant complex systematic reviews (T-CRCSRs, *vide supra*) (Chiappelli 2010; Chiappelli et al. 2010a, b). Osteoimmunology, in terms of the myriad of the fundamental cell biology and the intricate spectrum of clinical conditions it composes, appears to be a most likely candidate for the establishment of T-TRIP in the next decade.

3.2.3 Implications for Stomatology

Clearly, as we now look with a bird's eye view to the field of osteoimmunology and oestoimmunopathology, it becomes apparent that the increased understanding of the fundamental biology that underlies the complexly intertwined cross-interactions between bone and cellular immune metabolism will proffer substantial benefits in the treatment of a variety of ailments and diseases in the next decade. The implications and applications of these advancements in clinical osteoimmunoly, and specifically in translational evidence-based osteoimmunology will have broad pertinence both systemically and locally in certain anatomical or physiological domains and systems.

The head and neck region, and specifically its rostral aspect, which as noted earlier, is characterized by a complexity of bony structures, soft and hard mucosa where immune responses abound. In this anatomical context, which also houses one of the most complex joint of the body (i.e., the temporomandibular joint), fundamental understanding of osteoimmunological principles, pathways, and processes will permit a better grasp of stomatological pathobiology. That is to say, the elements of basic biology we have discussed above (*vide supra*, Chap. 1), which are essential to productive translational evidence-based osteoimmunology in general (*vide supra*, Chap. 2), apply to with equal validity to progress in oral biology and medicine in the coming years.

Case in point ought to be evident in the case of implants needed in aging patients with osteoporosis, or cancer patients with osteonecrosis of the mandible or the maxilla. Bone needs to be rebuilt in those cases, repacked with either (most often) palatal graphs, or with differentiated mineralizing osteoblastic cultures. In either situation, however, to ignore the immune dimension of the osteoimmune transaction inherent in the intervention will signify increased risk for clinical failure.

In patients with excessive lateral grinding, the temporomandibular joint may become impaired, consequentially to lateral stretching and pulling of the joint ligaments (*vide supra*), and masticatory musculature. This clinical situation, which is most often noted by temporomandibular joint specialists, and more rarely by the general dentist for the specific changes to the anatomical features primarily of the molars, can, if left unchecked or uncorrected, lead progressively to a bone-on-bone grinding situations, associated inflammation, and full blown osteoimmune pathology at the temporomandibular site. Certain patients, rather than grinding side-to-side, clench, often during sleep or unconsciously, with excessive force of the masticatory muscles (*vide supra*). Repeated action of this sort may lead to the molars, primarily, being pushed deeper into the mandibular and maxillary infraalveolar bones, such that the bone resorbs at these sites. The effect results in two serious clinical outcomes with osteoimmune implications. Firstly, inflammation will develop as bone tissue is harmed or destroyed in the process of progressive teeth pushing within the socket, as the infraalveolar bone is resorbed, and the tooth cementum undergoes osteoclastic resorption. Secondly, the joint is harmed because, the support provided by the molars is now diminished, which results in a shortening of the vertical dimension, a disordered functioning of the temporomandibular articulation.

The literature is mixed with respect to whether or not osteoporosis per se leads to clinically remarkable dysfunction of the temporomandibular joint. Osteoimmunology provides a useful paradigm that permits the elaboration of a sound mechanistic hypothesis. Stated simply, osteoclastic activity, which is characteristic in osteoporosis, may target the cartilaginous coruna that sits on the condyle. As osteoclastic activity resorbs this protective cartilage to a sufficient extent, the risk of bone-on-bone grinding increases with every movement of the joint. Inflammation will ensue, which will engender more active osteoclastic activity at the temporomamdibular joint through the processes described in preceding sections (*vide supra*).

In brief, osteoimmune events and processes are fundamental to normal and regenerative physiology systemically and locally in the oral cavity. Osteoimmune pathologies lead to significant impairments of structures and cell-mediated regulatory responses across anatomical regions determined by the axial, articular, and rostral skeletons. Comparative effectiveness analysis and evidence-based decision-making converge in the search for the best available clinical evidence, through the articulated research synthesis design and meta-analytical inferences, to procure the most efficacious and effective – in terms of raising clinical benefit while minimizing risk and costs – evidence-based osteoimmune interventions for oral and systemic medicine.

Chapter 4
Summary and Conclusion

We have attempted to present a framework for translational evidence-based osteoimmunology and osteoimmunopathology. We have established the biological foundations of the emerging understanding of the interactions between the bone system and the immune system. We reviewed the cellular, molecular, and proteomic pathways that converge to bone metabolism on one hand, and cell-mediated immunity on the other. More importantly, we described and discussed the multiple cross-regulatory mechanisms through which osteology and immunology sustain each other, and on each other.

A clear understanding of normal physiology is critical for the elucidation of pathological processes. Therefore, we built upon the physiological relations explicated in Part I in exploring a variety of pathologies that either are anchored in an osteoimmune etiology, or partake osteoimmune-driven events in their exacerbation. We noted that a common denominator of osteoimmune diseases is the fact that at their core lies either cellular immune dysfunction that engenders an alteration in the metabolic balance between osteoblastic bone generation and osteoclatic bone destruction, or a deregulation in bone metabolism that induces inflammatory and related cellular immune responses. Taken together, the observations at hand point to the fact that, should we obtain proteomic signatures of these, and other, osteoimmune pathologies, then interventions could be designed, tested, and evaluated both for efficacy (i.e., clinically relevant successful outcome – the "it's-working" experience), and effectiveness (i.e., the "it's safe, beneficial and affordable" experience).

Thus, we engaged in Part III of this writing, which established the foundations of the science of research synthesis, as it aims to retrieve, evaluate, and report the best available evidence for the efficacy or for the effectiveness of any given intervention under study. We, of course, focused on the research synthesis already available in the context of the osteoimmune pathologies described and discussed in Part II, and whose underlying fundamental biology could be elucidated based upon the elemental concepts explicated in Part I. That is to say, we attempted to lay the foundation of a novel perspective on clinical osteoimmunopathology: one that relies on translational evidence-based practice. To ensure that the approach we proposed remain cutting-edge, we explored and outlined the current and evolving parameters of evidence-based efficacy and effectiveness.

F. Chiappelli, *Osteoimmunopathology: Evidence-Based Perspectives from Molecular Biology to Systems Biology*, DOI 10.1007/978-1-4419-9495-0_4,

Increasingly, an intimate relationship has been recognized between stomatological and systemic physiopathological processes. Of course, this intimate intertwining of health and disease of the oral cavity and of the rest of the body pertains to, and involves as well the osteoimmune system. Thus, we have emphasized in these pages the relevance of oral biology and medicine to translational evidence-based osteoimmunology and osteoimmunopathology.

Glossary[1]

First pharyngeal arch	The first of six bulges or "arches" that develop in the human embryo about gestational week 4–6, is also called the mandibular arch. It divides into a maxillary process to give rise to structures including the bones of the lower two-thirds of the face (i.e., maxilla) up to and including the incus and malleus of the middle ear, as a superior regression of the Meckel's cartilage (cf., Meckel cartilage); and a mandibular process that comes to form a cartilaginous "model" of the mandible as a cartilaginous bar of the mandibular arch known as cartilage of Meckel.
ADAMTS	A family of peptidases defined by carrying "A Disintegrin And Metalloproteinase with Thrombospondin Motifs."
AGREE	Appraisal of Guidelines Research and Evaluation-Europe.
AHRQ	Agency for Healthcare Research and Quality (ahrq.gov).
AIDS	Acquired immune deficiency syndrome.
AMSTAR & R-AMSTAR	Assessment of multiple systematic reviews instrument, and its revised form: a tool that permits the evaluation of research quality of systematic reviews.
ANCA	Anti-neutrophil cytoplasmic antibodies: a group of autoantibodies, mainly of the IgG type, often indicative of an autoimmune disease.
Ankylosing spondylitis	Chronic, inflammatory arthritis and autoimmune disease that affects principally the thoracic and the sacro-ilium vertebral joints, and may lead to eventual fusion of the spine.

[1] The interested reader is also referred to the *US Health Information Knowledgebase* page (ushik.ahrq.gov), and the *Vocabularies* link therein.

F. Chiappelli, *Osteoimmunopathology: Evidence-Based Perspectives from Molecular Biology to Systems Biology*, DOI 10.1007/978-1-4419-9495-0,

Apoptosis	Programmed cell death that involves characteristic cell and nuclear changes, and death in a well-orchestrated and programmed pattern of events (e.g., blebbing, loss of cell membrane asymmetry and attachment, cell shrinkage, nuclear fragmentation, chromatin condensation, and chromosomal DNA fragmentation). Distinct from necrotic cell death, a process by which a cell dies as a result of toxicity, stress, or other traumatic challenges.
ART	Antiretroviral therapy.
Arthrodial joint	Gliding, synovial joint.
Bechterew's disease	See Ankylosing spondylitis.
Best available evidence	Following the process of research synthesis, which emerges from a PICO question that generates the appropriate search and identification of all of the available evidence, the best available evidence results from a process of assessment of the level of the evidence and of the quality of the evidence, which leads to acceptable sampling analysis and meta-analysis. What emerges from the process is a consensus of the best available evidence.
BGLAP	Bone gamma-carboxyglutamic acid-containing protein, also referred to as osteocalcin.
Bisphosphonates	Class of drugs that have two phosphonate (PO_3) groups, and that act to prevent the loss of bone mass.
Bisphosphonates	Drugs with two phosphonate groups (PO_3) used to prevent the loss of bone mass in osteoporosis and similar diseases, because they inhibit bone resorption by osteoclasts.
BSP-1	Bone sialoprotein I, cf. osteopontin.
Calcitriol	See VitD: the hormonally active form of vitamin D with three hydroxyl groups (abbreviated 1,25-(OH)2D3 or simply 1,25(OH)2D). It increases the level of calcium (Ca^{2+}) in the blood by increasing the uptake of calcium from the gut into the blood, by decreasing the transfer of calcium from blood to the urine by the kidney, and increasing the release of calcium into the blood from bone.
CD	Cluster of differentiation.
CERT	Center for Education & Research on Therapeutics: an entity established and funded by AHRQ with the charge of conducting research and providing education with the goal of advancing the optimal utilization of treatment modalities, devices, technologies, and products, with full awareness of cost/benefit and risk/benefit considerations.

Comparative effectiveness analysis	Systematic analysis and comparison of treatment effectiveness, as defined as cost/benefit and risk/benefit among several treatment alternatives. The *Institute of Medicine* defined (2009) comparative effectiveness research and analysis as "the generation and synthesis of evidence that compares the benefits and harms of alternative methods to prevent, diagnose, treat, and monitor a clinical condition or to improve the delivery of care. The purpose of CER is to assist consumers, clinicians, purchasers, and policy makers to make informed decisions that will improve health care at both the individual and population levels."
CONSORT	Consolidated standard for randomized trials.
COX-2	Cyclooxygenase 2, an inducible prostaglandin-endoperoxide synthase 2 that acts both as dioxygenase and as a peroxidase.
CRCSR	cf., Clinically relevant complex systematic reviews.
DABAs	Dual action bone agents, which act by stimulating the proliferation of osteoblasts, and inhibiting the proliferation of osteoclasts.
Distraction osteogenesis	A surgical intervention used to reconstruct skeletal deformities primarily of the long bones, as well as deformities of the maxillary and mandibular bones.
DMARDs	Disease-modifying antirheumatic drugs.
EBPC	Evidence-Based Practice Center: an entity established and funded by AHRQ with the charge of: (1) developing evidence reports (cf., evidence reviews), (2) evaluate technology assessments, (3) assess relevance to clinical, social science/behavioral, economic, and related issues, with the goal of becoming "science partners" in an effort to improve the quality, effectiveness , and appropriateness of healthcare interventions.
EQUATOR	Enhancing the quality and transparency of health research: a collaborative international initiative dedicated to enhancing the quality and transparency of health research set-up specifically to advance high quality reporting of health research studies, and to promote good reporting practices including the wider implementation of reporting guidelines.
ETA-1	Early T-lymphocyte activation-1; cf., osteopontin.
Evidence reviews	Critical summaries of systematic reviews evidence reports.

Fauces	Two distinct mucous membrane structures that signify the posterior aspect of the oral cavity: anteriorly, the palatoglossal arch, and posteriorly, the palatopharyngeal arch. The fauces guard the organism from invasion of pathogens by means of the Waldeyer ring, named after the German anatomist Heinrich Wilhelm Gottfried von Waldeyer-Hartz (1836–1921), the ring of lymphoid tissue around and about the naso- and oro-pharynx (i.e., from superior to inferior, pharyngeal tonsils, Eustachian tubal tonsils, palatine tonsils lingual tonsils).
Gasserian nucleus	Also known as the trigeminal, or semilunar ganglion: a sensory ganglion of the trigeminal nerve of the Meckel's cavity within the dura mater, and that covers the trigeminal impression near the apex of the petrous part of the temporal bone, whence emerge the ophthalmic, maxillary, and mandibular branches of the trigeminal nerve, Cranial nerve V. Named after the Austrian anatomist Johann Lorenz Gasser (1723–1765).
Gingymal joint	A joint that moves primarily in one plane only, such as the temporomandibular joint.
gp120	HIV envelope protein, glycoprotein with a molecular weight of 120 kD. Gp120 is essential for virus entry into CD4+ cells as it seeks out the CD4 moiety on T lymphocytes, to which it binds by electrostatic, van der Waals interactions and hydrogen bonds, thereby permitting a port for entry of HIV into CD4+ cells. For obvious reasons, gp120 is the target of concerted research efforts toward the development of anti-HIV vaccines.
GRADE & Ex-GRADE	Grading Recommendations Assessment, Development and Evaluation, and its expanded revision: instrument that permits the identification of the best available evidence for reporting in systematic reviews, and for integration in revised clinical practice guidelines.
HAART	Highly active antiretroviral therapy.
Harvesian system of canals	Osteons together form the Harvesian system, named after the British physician and anatomist Clopton Havers [1657–1702]. Each osteon consists of concentric lamellae of bone matrix that surround the central Harvesian canal. These microanatomical structures form canal-like structures, the Haversian canals. Blood vessels and nerves course the canals, and thusly penetrate the bone tissue.

Health-Care Bill The Affordable Health Care for America Act (or HR 3962) was crafted by the United States House of Representatives in November 2009. On December 24, 2009, the Senate passed an alternative healthcare bill, the Patient Protection and Affordable Care Act (H.R. 3590), which became a law on March 23, 2010. It was shortly thereafter amended by the Health Care and Education Reconciliation Act of 2010 (H.R. 4872), which became a law on March 30, 2010. The law, signed by President Obama, provides for the phased introduction over 4 years of a comprehensive system of health insurance with reforms, aimed at ensuring health-care coverage for the large majority of US citizens.

HIV Human immunodeficiency virus.

Hyoid bone Rests at the level of the base of the mandible in the front and the third cervical vertebra behind, and provides attachment to the muscles of the floor of the mouth and the tongue above, the larynx below, and the epiglottis and pharynx behind.

IKK Inhibitor of NF-κB kinase.

Inferior nasal concha Extends horizontally along the lateral wall of the nasal cavity, posteriorly to articulate with the conchal crest of the palatine bone, anteriorly to articulate with the conchal crest of the maxilla.

Integrin The IκB kinase enzyme is part of the complex upstream of the NF-κB signal transduction cascade. It phosphorylates the inhibitor of κ B protein, and thus liberating it the NF-κB transcription factor, and unmasking its nuclear localization signals.

IRIS Immune reconstitution inflammatory syndrome: a spectrum of immunological abnormalities consequential to immune-directed therapies in HIV-seropositive & AIDS-afflicted patients. It is characteristically an immunopathology determined by a deregulated immune surveillance activity, and manifested as responses to a previously acquired opportunistic infection with an overwhelming inflammatory response that paradoxically makes the symptoms of infection worse. In IRIS, the CD4 count rapidly increases following overeffective immunotherapy, and the associated sudden increase in the inflammatory response produces nonspecific symptoms such as fever, and can in some cases exacerbate damage to the infected tissue. Patients are at risk for IRIS when infected with HIV, cytomegalovirus, herpes zoster, Mycobacterium avium complex, Pneumocystis pneumonia, or Mycobacterium tuberculosis. AIDS patients are at increased risk for IRIS when HAART is first initiated.

JNK c-Jun N-terminal kinase.

Malar Related to the cheekbones.

MALDI-TOF A variation of SELDI-TOF, where the protein or peptide sample is mixed with the matrix molecule in solution and small amounts of the mixture are deposited on a surface and allowed to dry. The sample and matrix co-crystallize as the solvent evaporates.

Mandible Articulates with the two temporal bones at the temporomandibular joints, complex synovial joints (i.e., gliding or arthrodial) whose motion, while considerable occurs in one plane only (i.e., hindge or gingymal), and which suffer from several dysfunctions, including inflammatory osteoarthritis, with significant local and systemic sequelae.

Maxilla Conists of the alveolar processes that hold the upper teeth, and its articulation laterally to the zygomatic bones, in addition to remarkable features, such as the zygomatic process, the frontal process, the palatine process, the infraorbital foramen, and the maxillary sinus.

Meckel's cartilage The cartilaginous bar of the mandibular arch, identified in 1821 and named after its discoverer, the German anatomist Johann Friedrich Meckel (1781–1833). The Meckel's cartilage is a mandibular process that comes to form a cartilaginous "model" of the mandible as a cartilaginous bar of the mandibular arch.

Meta-analysis Statistical protocol aimed at combining the outcomes of multiple homogeneous studies. Homogeneity (aka, consistency) is tested by the X^2-derived Cochran Q test or the I^2 test. Whereas, Cochran Q has low power in meta-analyses that combine few studies (n low), and excessive power when n is high, the I^2 test quantifies the amount of variation (i.e., heterogeneity, inconsistency in outcome) in results across studies beyond that expected by chance (i.e., akin to a coefficient of agreement). I^2 represents the percentage of total variation (i.e., akin to shared variance) in estimated effects across studies that is attributable to heterogeneity (i.e., lack of consistency) rather than chance alone. When n is low (i.e., low power), it is appropriate to report the extent of uncertainty associated with I^2 in the form of CI_{95}.

MMP Matrix metalloproteinases: a family of proteolytic enzymes with either zinc or cobalt conjugated with the protein structure in their active site at three characteristics sites usually involving histidine, glutamate, aspartate, lysine, or arginine as possible ligands. The fourth position is usually a labile water molecule. They may be either exo- or endopeptidases. Metalloendopeptidases are responsible for processing the extracellular matrix (hence, MMP nomenclature), and play an important role in cell migration and tumor metastasis.

NFAT	Nuclear Factor of Activated T cells.
NICE	National Institute for Health and Clinical Excellence: independent UK organization that aims at providing evidence-based and comparative effectiveness guidance for promoting good health and preventing and treating ill health.
ODF	Osteoclast differentiation factor. Cf, RANKL.
OPG, osteoprotegerin	A member of the TNF family of cytokines, which can inhibit the production of the bone resorbing cell population, the osteoclasts, by inhibiting the differentiation of osteoclast myeloid precursors derived from granulocyte/macrophage-forming colony units. Osteoprotegerin is a RANK homolog, and works by binding to RANKL, thus blocking RANK-RANKL signal. Of note in the context of translational evidence-based osteoimmunology is the observation that production of OPG is stimulated in vivo by estrogen, and by strontium ranelate, a drug used to combat osteoporosis and to build bone mineralization.
OPGL	Cf, RANKL.
ORCA	Organizational Readiness to Change Assessment: an instrument designed to assess and validate PARiSH.
Osteochondropathy	Disease of the bone and cartilage.
Osteoarthritis	Degenerative arthritis or degenerative joint disease, a group of mechanical abnormalities involving inflammation and consequential degradation of joints, articular cartilage, and subchondral bone.
Osteoblast	Fibroblast-like cells that are responsible for bone creation.
Osteocalcin	cf., BGLAP
Osteochondroma	Benign tumor of cartilage and bone.
Osteoclast	Myeloid-like cells responsible for bone degradation and resorption.
Osteoid	The unmineralized, organic portion of the bone matrix that forms prior to the maturation of bone tissue begun by osteoblasts.
Osteonecrosis	Cellular death of bone tissue (for example, due to lack of blood supply; avascular necrosis).
Osteon	Compact cortical bone is composed of cylindrical units of bone structure, the osteons, with a dimension of circa 0.2 mm in diameter.
Osteopontin	Extracellular SIBLING protein produced by osteoblasts, which acts both in bone regeneration and immune regulation processes.

Osteoporosis	Pathological condition of bone where the bone mineral density is reduced, bone microarchitecture is disrupted, and the amount and variety of proteins in bone is altered.
Palatine bone	Forms what is commonly known as the hard palate, and articulates with six bones of the deep face, important for the internal structure of the oral cavity: the sphenoid, ethmoid, maxilla, inferior nasal concha, vomer bones.
PARiSH	Promoting Action on Research Implementation in Health Services: a healthcare framework that is widely promoted to implement evidence-based clinical practices.
PI	Protease inhibitors.
PICO question	The initial step in conducting a systematic review involves setting the research question, which carefully defines the clinical problem and the patient population (P), the interventions (I) to be contrasted (C), and the clinical outcome (O) of interest.
PRISMA	Preferred Reporting Items for Systematic Reviews and Meta-analyses.
PTEN	Phosphatidylinositol-3 kinase (PI3K)-dependent Akt activation can be regulated through the tumor suppressor phosphatase and tensin homolog.
PTH	Parathyroid hormone.
PTH-rP	PTH-related protein.
QALY	Quality-Adjusted Life Years: a measure of disease burden, including both the quality and the quantity of life lived. QALY is based on the number of years of life that would be added by the intervention. Each year in perfect health is assigned the value of 1.0 down to a value of 0.0 for death, and years not be lived in full health are given a value between 0 and 1. This measure is common in utility-based decision-making, particularly in the context of incremental cost-effectiveness (ICER) ratios.
QUOROM	Quality of Reporting of Meta-analyses.
RANK	Receptor Activator of Nuclear Factor κβ, a type I membrane protein expressed by osteoclasts and dendritic cells, and involved in bone resorption and the facilitation of immune signaling.
RANKL	Receptor activator of the nuclear factor-κβ, ligand for Research Synthesis.
ROR	Retinoic acid-related orphan receptor.
SELDI-TOF	Surface-enhanced laser desorption/ionization (SELDI), time-of-flight (TOF) is an ionization method in mass spectrometry that is used for the analysis of protein mixtures for the purpose of the identification of specific diagnostic and prognostic biomarkers. The analysis is based upon spotting the protein mixture on a surface modified with a chemical functionality.

SERMs	Selective Estrogen Receptor Modulators.
Sharpey's fibers	Also known as bone fibers, or perforating fibers, they form a matrix of collagenous fibrous tissue that adheres periosteum to bone. The fibers are in the outer layer of periosteum, and penetrate into the outer circumferential and interstitial lamellae of bone tissue. In the maxilla and mandible, these fibers signify the terminal ends of the periodontal ligament that interconnect the cementum and the periosteum of the alveolar bone. They are named after the Scottish anatomist William Sharpey (1802–1880), who first described them in 1846.
SIBLING	Small integrin-binding ligand, N-linked glycoproteins, bone sialoprotein (60–80 kDa).
SOCS3	Suppressor of cytokine signaling family member 3.
SORT	Strength of Recommendation Taxonomy: important validated instrument to evaluate the quality of evidence and the strength of recommendations.
Splanchnocranium	i.e., Facial skeleton, also refererred to as viscerocranium.
Stat1	Signal Transducers and Activators of Transcription family member 1.
Stomatology	Specialty of medicine that addresses the mouth and its associated diseases, and their oral and systemic sequelae.
Stomodeum	Emerging mouth pit from the pericardium and the forming pharyngolaryngeal structures in the first pharyngeal arch. This structure will first arise as a depression between the developing structures that will become the brain and the pericardium.
STREGA	Standards for reporting of genetic association.
STRICTA	Standards for reporting interventions in clinical trials of acupuncture, and more generally, complementary and alternative medicine (e.g., Ayurvedic medicine).
STROBE	Standards for reporting of observational studies in epidemiology.
Synarthrodial joint	Immovable joints.
Systematic reviews	The published report of a research synthesis study.
T-bet	Retinoic acid-related orphan receptor (ROR)$\gamma\tau$.
T-CRCSR	Translational clinically relevant complex systematic reviews.
TACE	TNFα-converting enzyme, a metalloprotease.
TAK	TGF-β-inducible kinase.
TMD	Temporomandibular joint disorder.
TNF	Tumor necrosis factor-α.

TOF	Time-of-flight analysis used in SELDI and MALDI. TOF is a method of mass spectrometry analysis in which ions mass-to-charge ratio is determined via a time measurement. Ions are accelerated by an electric field of known strength. The acceleration produces an ion that has the same kinetic energy as any other ion with the same charge. The velocity of acceleration of the ion depends on its mass-to-charge ratio, and the time employed by the particle to reach the detector placed at a known distance is a function of the mass-to-charge ratio of the particle, such that lighter particles travel faster. These measurement of time traveled permit the computation of the ion's mass-to-charge ratio, based upon the known experimental parameters.
Tori	Bony protrusions in the palate or mandible.
TRAF	TNF-α Receptor-Associated Factor.
TRANCE	Tumor Necrosis Factor [TNF]-related activation-induced cytokine. Cf., RANKL.
TRIP	Process of translating research into practice.
TTRIP	The process that seeks the integration of translational research into the practice reality.
VitD	1α,25-dihydroxyvitamin D3.
Volkmann's canals	Volkmann's canals are named after the German physiologist and anatomist Alfred Wilhelm Volkmann (1800–1877). They constitute channels that are alternative to the Harvesian canals to provide a route for blood vessels and nerves to reach the principal osteonal canal and bring nutrients to the bone. Volkmann's canals cross-link together into a network that interconnects the Harvesian canals across different osteons.
Vomer bone	A thin, somewhat quadrilateral bone situated in the median plane that forms the hinder and lower part of the nasal septum. Along its two surfaces run the nasopalatine groove obliquely downward and forward to aid the nasopalatine nerve and vessels. Its inferior border articulates with the crest formed by the maxilla and palatine bones.
Waldeyer ring	See fauces.
Zygomatic bones	Play a critical structural role as they articulate with the maxilla, the temporal bone, the sphenoid bone, and the frontal bone. The malar aspect presents the zygomaticofacial foramen for the passage of the zygomaticofacial nerve and vessels; the temporal aspect supports articulation with the maxilla, and forms the anterior boundary of the temporal fossa, the lower a part of the infratemporal fossa, an irregularly shaped cavity, situated below and medial to the zygomatic arch, and bounded laterally by the ramus of mandible.

Notes

Activin-like kinase-2 (ALK2) is also referred to as Activin A receptor, type I (ACVR1) because it is encoded by the ACVR1 gene on chromosome 2q23-q24. It belongs to the transforming growth factor (TGF)-β superfamily of structurally related signaling proteins, and operate via transmembrane signals through a heteromeric complex of receptor serine kinases that can include at least two type I (I and IB), which determine the signaling events, and two type II (II and IIB) transmembrane receptors, which are responsible for binding the ligands and for regulating the expression of the type I receptors. Together, these receptors form a stable complex following binding of the ligand to the extracellular cysteine-rich domain. Formation of this complex induces phosphorylation with predicted serine/threonine specificity of type I receptors by type II receptors at the cytoplasmic domain (ten Dijke et al. 1993).

Apoptosis (Gk: apo: from; ptosis: falling) refers to the process of programmed cell death first described by the German cytologist Carl Christoph Vogt (1817–1895) in 1842, and further characterized in greater details morphologically in 1885 by Walther Flemming (1843–1905), the German cell biologist renown for his first characterization and description of chromatin 3 years earlier (1882). In 1965, the Australian electron microscopist, John Kerr distinguished apoptosis from necrotic cell death, a process by which a cell dies as a result of toxicity, stress, or other traumatic challenges in which the cellular debris can damage the organism. In contrast to necrosis, where cell death results from acute cellular injury, apoptosis, in general, confers advantages during an organism's life cycle. Sydney Brenner, Horvitz, and John Sulston shared the 2002 Nobel Prize in Medicine for their characterization of the process of apoptosis. Apoptosis entails a set of well-defined biochemical events that lead to characteristic cell and nuclear changes, and death in a well-orchestrated and programmed pattern of events (e.g., blebbing, loss of cell membrane asymmetry and attachment, cell shrinkage, nuclear fragmentation, chromatin condensation, and chromosomal DNA fragmentation).

Cluster of differentiation (CD) refers to membrane glycoproteins used for the identification and investigation of cell surface molecules expressed on white blood cells (leukocytes), originally. In more current usage, CD molecules are recognized on stem cells that may differentiate into cells other than immune cells, and, by extension, in differentiated cells of other systems. The membrane glycoproteins that are identified as CD are endowed with a variety of functions, including receptors, coreceptors, or ligands. The signal cascade that is initiated at the CD alters cell adhesion and migration properties, activation, proliferation, maturation, or programmed cell death (=apoptosis). The CD nomenclature was established at the first International Workshop and Conference on Human Leukocyte Differentiation Antigens (Paris 1982). As of now, the ninth International Conference on Human Leukocyte Differentiation Antigens-HLDA9 (Barcelona 2010), the number of identified CD has risen to 363 (CD363, sphingosine-1-phosphate receptor 1). The CD system is commonly used as cell markers to define cell populations based on function, degree of maturation, extent of activation, etc. CD34, for example, is a cell surface glycoprotein that functions as a cell–cell adhesion factor, and mediates the attachment of stem cells to bone marrow extracellular matrix or directly to stromal cells. CD34 is expressed on early hematopoietic and vascular-associated cells, and acts as an important adhesion molecule required for T cells to enter lymph nodes. In that sense, CD34 is one example of a critical CD glycoprotein that bridges bone and immune metabolism in the osteoimmune context discussed here.

Comparative effectiveness analysis is a process first developed by mathematicians and programmers to help those individuals make decisions for allocation of benefits, when markets and price signals were all but inaccessible. The principal use of comparative effectiveness analysis in that context was to guide decision-makers in making efficient allocation of goods and services. That is the reason why, when applied to health care, comparative effectiveness analysis is based on the assessment of the incremental cost and effects that result from choosing this vs. that strategic option (e.g., early childhood vaccination vs. no vaccination in light of reported rise in autism incidence). Thus, the purpose of comparative effectiveness analysis in health care is to assist the clinical decision-maker in determining how to allocate resources and services across competing needs to maximize beneficial health outcomes within a constrained budget. Often, decisions are aided by transformations of incremental costs per incremental quality-adjusted life years (QALYs[2]), which incorporate changes in both length and quality of life. Cost-to-QALY ratios estimate the extra cost required to achieve one additional quality-adjusted life year. In brief, a full comparative effectiveness analysis examines costs relative to quality of life, based on the allocation of a fixed budget across the competing interventions, and yields insight into the relative economic attractiveness of a given therapy, technology, or product. Whereas, it is incapable of incorporating societal value judgment or personal assessments (e.g., patient satisfaction), it is useful because it successfully disaggregates cost-consequence issues (Detsky and Laupacis 2007). As part of the 2009 American Recovery & Reinvestment Act (ARRA, PL-111-5), $1.1B were allocated for comparative effectiveness research & analysis.

[2] www.nice.org.uk/newsroom/features/measuringeffectivenessandcosteffectivenesstheqaly.jsp.

One hundred topics of urgent need were identified. Within the first quartile, one finds inflammatory diseases, including rheumatoid arthritis. Within the second quartile, one finds issues relating to the use of biphosphonates, osteopenia, vertebral fractiures, etc. Within the ultimate quartile of relevance and urgency, one finds studies of osteoarthritis.

Metalloproteases constitute a family of proteolytic enzymes, which are endowed, as the most prominent functional group in their active site, of either zinc or, less often cobalt, conjugated with the protein structure at three characteristics sites usually involving histidine, glutamate, aspartate, lysine, or arginine as possible ligands, and utilizing the fourth coordination position as a labile water molecule. Metalloproteases are generally either exopeptidases (EC 3.4.17) (i.e., metallocarboxypeptidases), or endopeptidases (EC 3.4.24). Metalloendopeptidases are responsible for processing the extracellular matrix, and thus play an important role in cell migration (e.g., lymphocyte homing), and tumor metastasis. Specifically, matrix metalloproteinases (MMPs) are predominantly zinc-dependent endopeptidases, and belong to the larger family of proteases known as the metzincin superfamily. They degrade extracellular matrix proteins, but also can process bioactive molecules. They can cleave cell surface receptors for the release of apoptotic ligands (e.g., FAS ligand), chemokines, cytokines, and growth factors. Hence, MMPs can regulate cell behavior (e.g., activation, proliferation, migration, differentiation, apoptosis), as well tissue and system performance (e.g., angiogenesis, immune defense, bone remodeling) (Bovée et al. 2010).

The *major histocompatibility complex* (MHC) is a large genomic region on chromosome 6 in humans that encodes MHC molecules, and defines "self" in terms of the immune surveillance processes. Stated simply, MHC molecules inside the cell take a biochemical fragment (i.e., "nonself") into a groove defined by the MHC tertiary structure, and display the fragment on the cell surface in a manner that is read by the immune system as "self+nonself." T cells are educated in the thymus to recognize "self" and to distinguish "self" from "self+nonself"; thus, T cells, under normal conditions, respond to presentation of "self+nonself" by a concerted activation process that, when productive, leads to T cell proliferation (i.e., clonal expansion), and maturation (i.e., from a virgin/naive T cell [CD4/CD8+CD45RA+] to a memory T cell [CD4/CD8+CD45R0+]) capable of specific cytokine responses (e.g., TH1, TH2) and engagement of specific pathways of immune surveillance responses. There are two principal classes of MHC molecules: MHC Class I is found on all cells and present elements of virally infected cells and tumor cells (i.e., "endogenous nonself" to CD8 T cells, whose function is predominantly cytotoxic – that is, removal of virally infected and tumor cells. MHC Class II is expressed on cells that function as antigen-presenting cells (APCs), which are primarily of myeloid origin (i.e., monocytes and their activated counterparts, macrophages; microglia in the brain parenchyma; dendritic cells). It is important to recall the shared myeloid origin of APCs and osteoclasts, in the context of this writing. Upon invasion by a pathogen, that foreign organism is processed by APCs, and fragments thereof associate with MHC Class II components intracytoplasmically.

These "self+nonself" complexes are translocated to the plasma mambrane, where they are presented to CD4 T cell, which primarily acts as helper T lymphocytes by engaging and promoting an antigen-dependent immune response. A third class of MHC has been described, MHC Class III, which includes genes coding several components of the complement system (e.g., C2, C4) and pro-inflammatory or heat-shock molecules. Whereas, MHC Class III clearly does not function similarly as either MHC Class I or II; it also sits on chromosome 6, and for this reason they both are frequently described together.

Osteopontin, aka bone sialoprotein I (BSP-1), or the early T-lymphocyte activation (ETA-1), or secreted phosphoprotein 1, is a 33-kDa SIBLING glycoprotein encoded by SPP1 on chromosome 4. In bone remodeling, it mediates anchoring osteoclasts to the mineral matrix of bones, and serves to initiate the process by which osteoclasts develop their ruffled borders to begin bone resorption (Merry et al. 1993; Choi et al. 2008). In the immune system, it is a pleiotropic cytokine expressed by activated T cells, dendritic cells, and macrophages and is upregulated during inflammation (Morimoto et al. 2010). It binds to several integrin receptors (e.g., $\alpha 4\beta 1$, $\alpha 9\beta 1$, and $\alpha 9\beta 4$) expressed by leukocytes, thus mediating immune cell adhesion and migration, in addition to immune cell activation, cytokine production, and immune cell survival by regulating apoptosis (Wang and Denhardt 2008). It is considered at this juncture a critical osteoimmune factor.

RANK signaling cascade is initiated when RANKL binds to the extracellular domain of RANK, which passes the signal along to TRAF6 (TNF receptor-associated factor 6). TRAF6 has various downstream mediators, which control the expression of osteoclast-specific genes during differentiation and activation of osteoclasts. AP-1 is activated by signaling cascades mediated by JNK (c-Jun N-terminal kinase), and the phosphorylation of the inhibitor of NF-κB kinase (IKK) which leads to the activation of NF-κB. Related cascades of mitogen-activated protein kinases such as the TGF-β-inducible kinase TAK1 and the p38 stress kinase participate in RANK signal transduction. p38 is activated via the phosphorylation by MKK6 and in turn activates the transcriptional regulator mi/Mitf, which is responsible for the transcriptional control of genes encoding for the osteoclast-specific enzymes TRAP and cathepsin K. ERK (extracellular signal-related kinase), a downstream target of MEK1, acts as a negative regulator of osteoclastogenesis for ERK inhibitors that have shown to accelerate RANKL-induced osteoclastogenesis. The serine/threonine kinase Akt and the phosphatidylinositol-3-OH kinase (PI(3)K) are downstream elements of src and are known to mediate cell survival, motility, and cytoskeletal rearrangements by the activation of the MEK/ERK and Akt/NFκB pathways. AFX/FOXO4 also contributes to modulate osteoclast survival (Colucci et al. 2004; Takayanagi 2007, 2009, 2010). Moreover, the association of paired immunoglobulin-like receptor A (PIR-A) and OSCAR to FcRγ and triggering receptor expressed on myeloid cells (TREM) 2 and signal-regulatory protein β1 (SIRPβ1) to DAP12 in osteoclast precursors is costimulatory signal for RANKL, since one signal by its own is not able to induce osteoclastogenesis (Colonna et al. 2007).

Vitamin D_3, here referred to as VitD, is one member of the vitamin D family[3] of fat-soluble secosteroids, such that vitamin D without a subscript refers to either D_2 or D_3 or both. Vitamin D_1 is the molecular compound of ergocalciferol with lumisterol, in equi-molar ratio. Vitamin D_4 is 22-dihydroergocalciferol, and Vitamin D_5 is sitocalciferol. Again, here we have used the consistent abbreviation VitD to indicate vitamin D_3 (i.e., cholecalciferol, a derivative of cholesterol), which results in the skin, from the ultraviolet irradiation of 7-dehydrocholesterol. Production is greatest in the stratum basale (=inner most) and stratum spinosum (=inner) layers of the epidermal strata of the skin. Whether produced as described in the skin epidermis, or ingested, cholecalciferol is hydroxylated by hepatic microsomal enzyme vitamin D-25-hydroxylase at position 25 to form calcidiol (aka, 25-hydroxycholecalciferol), which is in turn metabolized to the active VitD form, calcitriol, by 25-hydroxyvitamin D_3 1-alpha-hydroxylase, itself activated by PTH, low calcium and low phosphate levels. Calcitriol is then released in the circulation, where it binds to its carrier chaperon protein, vitamin D-binding protein (VDBP), which aids its binding to the vitamin D receptor at appropiate targets. The receptor, a member of the superfamily of steroid/thyroid hormone receptors, is primarily localized in the nucleus, and has a finely controlled translocation life cycle. Binding of calcitriol to the receptor engages transcription regulation that modulates the gene expression of transport proteins to increase calcium absorption. Calcitriol engagement of this receptor also regulates cell proliferation and differentiation of a variety of cell populations, including myeloid precursors, whence osteoclasts derive, and T lymphocytes, thus underscoring the important role of VitD in the osteoimmune intertwined network of cross-regulatory processes and events.

Vladimir Mikhailovich Bekhterev (Bechterev) (20 January 1857– 24 December 1927), Russian Neurologist. He first described Ankylosing Spondylitis or Bekhterev's disease while conducting research at the University of Kazan, where he was the head of the Psychiatry Department (1885–1893). Note: the name of the disease became Bechterew's disease in English, following the German transliteration system for Russian names. Interestingly, as he was a most renowned psychiatrist of his time, he was called by Stalin for a consult. On the Day before Christmas 1927, Bekhterev diagnosed Stalin with "grave paranoia," and later that same day Bekhterev suddenly died.

[3] So named in 1921 by American biochemist Elmer Verner McCollum (1879–1967).

References

Abdallah BM, Stilgren LS, Nissen N, Kassem M, Jorgensen HR, Abrahamsen B. Increased RANKL/OPG mRNA ratio in iliac bone biopsies from women with hip fractures. Calcif Tissue Int 2005 76:90–7.

Aggarwal S, Gurney AL. IL-17: prototype member of an emerging cytokine family. J. Leukoc. Biol. 2002 71:1–8.

Aghaloo TL, Chaichanasakul T, Bezouglaia O, Kang B, Franco R, Dry SM, Atti E, Tetradis S. Osteogenic potential of mandibular vs. long-bone marrow stromal cells. J Dent Res. 2010 89:1293–8.

AGREE Collaboration. Development and validation of an international appraisal instrument for assessing the quality of clinical practice guidelines: the AGREE project. Qual Saf Health Care 2003 12:18–23.

Ajaj RA, Barkhordarian A, Phi L, Giroux A, Chiappelli F. Evidence-Based Dentistry: The Next Frontier in Translational and Trans-National Dental Practice. Dental Hypotheses. 2011 2:55–62.

Alagoz O, Hsu H, Schaefer AJ, Roberts MS. Markov decision processes: a tool for sequential decision making under uncertainty. Med Decis Making. 2010 30:474–83.

Anderson GI, MacQuarrie R, Osinga C, Chen YF, Langman M, Gilbert R. Inhibition of leukotriene function can modulate particulate-induced changes in bone cell differentiation and activity. J Biomed Mater Res 2001, 58:406–11.

Andronis L, Barton P, Bryan S. Sensitivity analysis in economic evaluation: An audit of NICE current practice and a review of its use and value in decision-making. Health Technol Assess. 2009 iii, ix-xi: 1–61.

Angeli A, Dovio A, Sartori ML, Masera RG, Ceoloni B, Prolo P, Racca S, Chiappelli F. Interactions between glucocorticoids and cytokines in the bone microenvironment. Annals of the New York Academy of Sciences, 2002, 966:97–107.

Annunziato F, Cosmi L, Santarlasci V, Maggi L, Liotta F, Mazzinghi B, Parente E, Filì L, Ferri S, Frosali F, Giudici F, Romagnani P, Parronchi P, Tonelli F, Maggi E, Romagnani S. Phenotypic and functional features of human Th17 cells. J Exp Med. 2007 204:1849–61.

Antonelli LRV, Mahnke Y, Hodge JN, Porter BO, Barber DL, DerSimonian R, Greenwald JH, Roby G, Mican JA, Sher A, Roederer M, Sereti I. Elevated frequencies of highly activated CD4+ T cells in HIV+ patients developing immune reconstitution inflammatory syndrome. Blood, 2010, 116:3818–27.

Arora S, Agrawal M, Sun L, Duffoo F, Zaidi M, Iqbal J. HIV and bone loss. Curr Osteoporos Rep. 2010 8:219–26.

Arrich J, Piribauer F, Mad P, Schmid D, Klaushofer K, Müllner M. Intra-articular hyaluronic acid for the treatment of osteoarthritis of the knee: systematic review and meta-analysis. CMAJ 2005 172:1039–43.

Arron JR, Choi Y. Osteoimmunology: Bone versus immune system. Nature 2001 408:535–6.

Atieh MA, Ibrahim HM, Atieh AH. Platform Switching for Marginal Bone Preservation Around Dental Implants: A Systematic Review and Meta-Analysis. J Periodontol. 2010 81(10): 1350–66.

Baio G, Dawid AP. Probabilistic Sensitivity Analysis in Health Economics. Research Report No. 292, Department of Statistical Science, University College London, March 2008.

Barkhordarian A, Ajaj R, Ramchandani MH, Demerjian G, Cayabyab R, Danaie S, Ghodousi N, Iyer N, Mahanian N, Phi L, Giroux A, Manfrini E, Cajulis OS, Brant XM, Chiappelli F. Osteoimmunopathology in HIV/AIDS: A Translational Evidence-Based Perspective. Pathology Research International, 2011 (In Press).

Barkhordarian A, Iyer N, Shapshak P, Somboonwit C, Sinnot J, Chiappelli F. Cellular Immune-Mediated Surveillance Modulated by TH17 & Tregs. Bioinformation. 2011 6:39–40.

Barteczko K, Jacob M. A re-evaluation of the premaxillary bone in humans. Anat Embryol (Berl). 2004 207:417–37.

Bartolucci AA, Hillegas WB. Overview, Strengths, and Limitations of Systematic Reviews and Meta-Analyses. In Chiappelli F. Brant XMC, Oluwadara OO, Neagos N, Ramchandani MH. (Eds.) Understanding Evidence-Based Practice: Toward Optimizing Clinical Outcomes. Springer, Heidelberg 2010, Chap 2.

Begg C, Cho M, Eastwood S, Horton R, Moher D, Olkin I, Pitkin R, Rennie D, Schulz KF, Simel D, Stroup DF. Improving the quality of reporting of randomized controlled trials. The CONSORT statement. JAMA. 1996 276:637–9.

Bellamy N, Campbell J, Robinson V, Gee T, Bourne R, Wells G. Viscosupplementation for the treatment of osteoarthritis of the knee. Cochrane Database Syst Rev. 2005 18:CD005321.

Bernstein M, Kovar H, Paulussen M, et al. Ewing sarcoma family of tumors: current management. Oncologist 2006 11: 503–19.

Bianco P, Kuznetsov SA, Riminucci M, Robey PG. Postnatal skeletal stem cells. Methods Enzymol. 2006 419:117–48.

Bolland MJ, Grey AB, Gamble GD, Reid IR. Effect of osteoporosis treatment on mortality: a meta-analysis. J Clin Endocrinol Metab. 2010 95:1174–81.

Bond TE Jr. *A practical treatise on dental medicine*. Philadelphia: Lindsay & Blakiston, 1848.

Bonjoch A, Figueras M, Estany C, Perez-Alvarez N, Rosales J, Del Rio L, di Gregorio S, Puig J, Gómez G, Clotet B, Negredo E; the Osteoporosis Study Group. High prevalence of and progression to low bone mineral density in HIV-infected patients: a longitudinal cohort study. AIDS. 2010 24:2827–33.

Boonen S, Laan RF, Barton IP, Watts NB. Effect of osteoporosis treatments on risk of non-vertebral fractures: review and meta-analysis of intention-to-treat studies. Osteoporos Int. 2005 16:1291–8.

Bordin L, Priante G, Musacchio E, Giunco S, Tibaldi E, Clari G, Baggio B. Arachidonic acid-induced IL-6 expression is mediated by PKCα activation in osteoblastic cells. Biochemistry 2003 42:4485–91.

Bovée JV. Multiple osteochondromas. Orphanet J Rare Dis. 2008 3:3.

Bovée JV, Hogendoorn PC, Wunder JS, Alman BA. Cartilage tumours and bone development: molecular pathology and possible therapeutic targets. Nat Rev Cancer. 2010 10:481–8.

Boyce BF, Xing L. Biology of RANK, RANKL, and osteoprotegerin. Arthritis Res. Ther. 2007, 9S: 1–27.

Breslau ES, Rochester PW, Saslow D, Crocoll CE, Johnson LE, Vinson CA. Developing partnerships to reduce disparities in cancer screening. Prev Chronic Dis. 2010 7:A62.

Brinkrolf P, Landmeier S, Altvater B, Chen C, Pscherer S, Rosemann A, Ranft A, Dirksen U, Juergens H, Rossig C. A high proportion of bone marrow T cells with regulatory phenotype (CD4+CD25hiFoxP3+) in Ewing sarcoma patients is associated with metastatic disease. Int J Cancer. 2009 125:879–86.

Brouwers MC, Kho ME, Browman GP, Burgers JS, Cluzeau F, Feder G, Fervers B, Graham ID, Hanna SE, Makarski J. AGREE Next Steps Consortium. Development of the AGREE II, part 1: performance, usefulness and areas for improvement. CMAJ. 2010a 182:1045–52

Brouwers MC, Kho ME, Browman GP, Burgers JS, Cluzeau F, Feder G, Fervers B, Graham ID, Hanna SE, Makarski J. AGREE Next Steps Consortium. Development of the AGREE II, part 2: assessment of validity of items and tools to support application. CMAJ. 2010b 182:E472–8.

Brown TT, Qaqish RB. Antiretroviral therapy and the prevalence of osteopenia and osteoporosis: a meta-analytic review. AIDS. 2006 20:2165–74.

Brozek JL, Akl EA, Jaeschke R, Lang DM, Bossuyt P, Glasziou P, Helfand M, Ueffing E, Alonso-Coello P, Meerpohl J, Phillips B, Horvath AR, Bousquet J, Guyatt GH, Schünemann HJ. GRADE Working Group. Grading quality of evidence and strength of recommendations in clinical practice guidelines: Part 2 of 3. The GRADE approach to grading quality of evidence about diagnostic tests and strategies. Allergy. 2009 64:1109–16.

Brzusek D, Petron D. Treating knee osteoarthritis with intra-articular hyaluronans. Curr Med Res Opin. 2008 24:3307–22.

Burkiewicz JS, Scarpace SL, Bruce SP. Denosumab in osteoporosis and oncology. Ann Pharmacother. 2009 43:1445–55.

Caetano-Lopes J, Canhão H, Fonseca JE. Osteoblasts and bone formation. Acta Reumatol Port. 2007 32:103–10.

Carter PH, Schipani E. The roles of parathyroid hormone and calcitonin in bone remodeling: prospects for novel therapeutics. Endocr Metab Immune Disord Drug Targets. 2006, 6:59–76.

Cazanave C, Dupon M, Lavignolle-Aurillac V, Barthe N, Lawson-Ayayi S, Mehsen N, Mercié P, Morlat P, Thiébaut R, Dabis F; Groupe d'Epidémiologie Clinique du SIDA en Aquitaine. Reduced bone mineral density in HIV-infected patients: prevalence and associated factors. AIDS. 2008 22:395–402.

Chang EJ, Kim HJ, Ha J, Kim HJ, Ryu J, Park KH, Kim UH, Lee ZH, Kim HM, Fisher DE, Kim HH. Hyaluronan inhibits osteoclast differentiation via Toll-like receptor 4. J Cell Sci. 2007 120:166–76.

Chiappelli F, Ramchandani MH, Phi L, Brant XMC. Clinically Relevant Complex Systematic Reviews in Endodontics: Relevance to Comparative Effectiveness Research and Evidence-Based Clinical Decision-Making. In F. Chiappelli & Brant XMC (Eds.) Comparative Effectiveness Research and Evidence-Based Health Care in Endodontics. Bentham Science, E-books, 2010, Chapter 1.

Chiappelli F. Future Avenues of Research Synthesis for Evidence-Based Clinical Decision Making. In Chiappelli F. Brant XMC, Oluwadara OO, Neagos N, Ramchandani MH. (Eds.) Understanding Evidence-Based Practice: Toward Optimizing Clinical Outcomes. Springer, Heidelberg 2010, Chap 15.

Chiappelli F. The Science of Research Synthesis: A Manual of Evidence-Based Research for the Health Sciences – Implications and Applications in Dentistry. NovaScience Publisher, Inc., 2008. pp. 1–327.

Chiappelli F, Cajulis OS. The logic model in evidence-based clinical decision-making in dental practice. J Evid Based Dent Pract 2009 9:206–10.

Chiappelli F, Cajulis O, Newman M. Comparative Effectiveness Research in Evidence Based Dental Practice. J Evid Based Dent Pract 2009 9:57–8.

Chiappelli F, Avila J, Chai DY, Dimaano C, Kudaverdian A, Park SH, Peng S, Sahhal S. Evidence-Dentistry in the XXI Century: Toward Clinically Relevant Complex Systematic Reviews (CRCSRs). The Open Dentistry Journal 2010 4:1–3.

Choi Y. Advances of Experimental Biology and Medicine: Osteoimmunology: Interactions of the Immune and Skeletal Systems II; Springer, 2009

Choi, Y, Ed. Osteoimmunology: RANKL Inhibition: From Mice to Men [and Women]; 2007

Choi ST, Kim JH, Kang EJ, Lee SW, Park MC, Park YB, Lee SK. Osteopontin might be involved in bone remodelling rather than in inflammation in ankylosing spondylitis. Rheumatology 2008, 47:1775–9.

Clay PG, Voss LE, Williams C, Daume EC. Valid treatment options for osteoporosis and osteopenia in HIV-infected persons. Ann Pharmacother. 2008 42:670–9.

Cohen MM. The New Bone Biology: Pathologic, Molecular, and Clinical Correlates. Am J Med Genet 2006, 140A:2646–706

Colonna M, Turnbull I, Klesney-Tait J. The enigmatic function of TREM-2 in osteoclastogenesis. Adv Exp Med Biol. 2007 602:97–105

Colucci S, Brunetti G, Rizzi R, Zonno A, Mori G, Colaianni G, Del Prete D, Faccio R, Liso A, Capalbo S, Liso V, Zallone A, Grano M. T cells support osteoclastogenesis in an in vitro model derived from human multiple myeloma bone disease: the role of the OPG/TRAIL interaction. Blood. 2004 104:3722–30.

Dai J, He P, Chen X, Greenfield E. TNFα and PTH utilize distinct mechanisms to induce IL-6 and RANKL expression with markedly different kinetics. Bone 2006 38:509–20.

Dardenne O, Prud'homme J, Arabian A, Glorieux R, St-Arnaud R. Targeted inactivation of the 25-hydroxyvitamin D(3)-1(alpha)-hydroxylase gene (CYP27B1) creates an animal model of pseudovitamin D-deficiency rickets. Endocrinology 2001 142:3135–41.

Das Sarma J. γδ T cells and IL-17/IL-17R signaling axis in CNS inflammation. Int J Interferon, Cytokine Mediator Res 2010 2:149–151.

Day AJ, de la Motte CA. Hyaluronan cross-linking: a protective mechanism in inflammation? Trends Immunol. 2005 26:637–43.

de la Motte C, Nigro J, Vasanji A, Rho H, Kessler S, Bandyopadhyay S, Danese S, Fiocchi C, Stern R. Platelet-derived hyaluronidase 2 cleaves hyaluronan into fragments that trigger monocyte-mediated production of proinflammatory cytokines. Am J Pathol. 2009 174:2254–64.

Demerjian GG, Sims AB, Stack BC. Proteomic Signature of Temporomandibular Joint Disorders (TMD): Toward Diagnostically Predictive Biomarkers. Bioinformation, 2010 – In Press

Detsky AS, Laupacis A. Relevance of cost-effectiveness analysis to clinicians and policy makers. JAMA 2007 298:221–4.

Dittus RS, Roberts SD, Wilson JR. Quantifying uncertainty in medical decisions. J Am Coll Cardiol. 1989 14:S23A–8A.

Donaldson NE, Rutledge DN, Ashley J. Outcomes of adoption: measuring evidence uptake by individuals and organizations. Worldviews Evid Based Nurs. 2004 1S:41–51.

Dougall WC, Glaccum M, Charrier K, Rohrbach K, Brasel K, De Smedt T, Daro E, Smith J, Tometsko ME, Maliszewski CR, Armstrong A, Shen V, Bain S, Cosman D, Anderson D, Morrissey PJ, Peschon JJ, Schuh J. RANK is essential for osteoclast and lymph node development. Genes Dev 1999, 13:2412–24.

Ducy P, Zhang R, Geoffroy V, Ridall AL, Karsenty G. Osf2/cbfa1: A transcriptional activator of osteoblast differentiation. Cell 1997 89:747–54.

Dunn GF, Hack GD, Robinson WL, Koritzer RT. Anatomical observation of a craniomandibular muscle originating from the skull base: the sphenomandibularis. *Cranio* 1996, 14:97–103

Eghbali-Fatourechi G, Khosla S, Sanyal A, Boyle W, Lacey D, Riggs B: Role of RANK ligand in mediating increased bone resorption in early postmenopausal women. J Clin Invest 2003 111:1221–30.

Escalas C, Trijau S, Dougados M. Evaluation of the treatment effect of NSAIDs/TNF blockers according to different domains in ankylosing spondylitis: results of a meta-analysis. Rheumatology (Oxford). 2010 49:1317–25.

Ewing J. Neoplastic Diseases: A Textbook on Tumors. WB Saunders, Philadelphia, 1919

Faggion CM Jr. Grading the quality of evidence and the strength of recommendations in clinical dentistry: a critical review of 2 prominent approaches. J Evid Based Dent Pract. 2010 10:78–85.

Fahrleitner-Pammer A, Dobnig H, Piswanger-Soelkner C, Bonelli C, Dimai HP, Leb G, Obermayer-Pietsch B. Osteoprotegerin serum levels in women: correlation with age, bone mass, bone turnover and fracture status. Wien Klin Wochenschr 2003 115:291–7.

Fakruddin JM, Laurence J. HIV envelope gp120-mediated regulation of osteoclastogenesis via receptor activator of nuclear factor kappa B ligand (RANKL) secretion and its modulation by certain HIV protease inhibitors through interferon-gamma/RANKL cross-talk. J Biol Chem. 2003 278:48251–8.

Fox SW, Lovibond AC. Current insights into the role of transforming growth factor-beta in bone resorption. Mol Cell Endocrinol 2005 243:19–26.

Fukunaga M, Sone T.Blood concentrations and secretory capacity of calcitonin in blood in osteoporosis. Clin Calcium. 2001 11:1140–5.

Gaffen SL. Biology of recently discovered cytokines: interleukin-17 – a unique inflammatory cytokine with roles in bone biology and arthritis. Arthritis Res Ther. 2004 6:240–7.

Gallois A, Mazzorana M, Vacher J, Jurdic P. Osteoimmunology and the effects of the immune system on bone. Med Sci [Paris]. 2009, 25:259–65.

Gliklich RE, Dreyer NA, Eds. Registries for Evaluating Patient Outcomes: A User's Guide. 2nd ed. (Prepared by Outcome DEcIDE Center [Outcome Sciences, Inc. d/b/a Outcome] under Contract HHSA29020050035I TO3.) AHRQ Pub. 10-EHC049. Rockville, MD: Agency for Healthcare Research and Quality. September 2010.

Goeree R, Blackhouse G, Adachi J. Cost-effectiveness of alternative treatments for women with osteoporosis in Canada. Curr Med Res Opin. 2006 22:1425–36.

Graudal N, Jürgens G. Similar effects of disease modifying anti rheumatic drugs, glucocorticoids and biologics on radiographic progression in rheumatoid arthritis: meta-analysis of 70 randomized placebo or drug controlled studies including 112 comparisons. Arthritis Rheum. 2010 62(10):2852–63.

Grcević D, Lee SK, Marusić A, Lorenzo JA. Depletion of CD4 and CD8 T lymphocytes in mice in vivo enhances Vit Dxyvitamin D3-stimulated osteoclast-like cell formation in vitro by a mechanism that is dependent on prostaglandin synthesis. J Immunol. 2000 165:4231–8.

Greco SJ, Corcoran KE, Cho KJ, Rameshwar P. Tachykinins in the emerging immune system: relevance to bone marrow homeostasis and maintenance of hematopoietic stem cells. Front Biosci. 2004, 9:1782–93.

Green J, Clézardin P. The molecular basis of bisphosphonate activity: a preclinical perspective. Semin Oncol. 2010 37:S3–11

Griebsch I, Coast J, Brown J. Quality-adjusted life-years lack quality in pediatric care: a critical review of published cost-utility studies in child health. Pediatrics. 2005 115:e600–14.

Grieve R, Nixon R, Thompson SG. Bayesian Hierarchical Models for Cost-Effectiveness Analyses that Use Data from Cluster Randomized Trials. Med Decis Making 2010 30:163–75

Grigrie D, Neacsu E, Marinescu M, Popa O. Circulating osteoprotegerin and leptin levels in postmenopausal women with and without osteoporosis. Rom J Intern Med 2003 41:409–15.

Grijsen ML, Vrouenraets SM, Steingrover R, Lips P, Reiss P, Wit FW, Prins JM. High prevalence of reduced bone mineral density in primary HIV-1-infected men. AIDS. 2010 24:2233–8.

Grund B, Peng G, Gibert CL, Hoy JF, Isaksson RL, Shlay JC, Martinez E, Reiss P, Visnegarwala F, Carr AD; INSIGHT SMART Body Composition Substudy Group. Continuous antiretroviral therapy decreases bone mineral density. AIDS. 2009 23:1519–29.

Gu G, Mulari M, Peng Z, Hentunen T, Väänänen HK. Death of osteocytes turns off the inhibition of osteoclasts and triggers local bone resorption. Biochem Biophys Res Commun. 2005 335:1095–101.

Guaraldi G, Ventura P, Albuzza M, Orlando G, Bedini A, Amorico G, Esposito R. Pathological fractures in AIDS patients with osteopenia and osteoporosis induced by antiretroviral therapy. AIDS. 2001 15:137–8.

Ha H, Lee JH, Kim HN, Kim HM, Kwak HB, Lee S, Kim HH, Lee ZH. alpha-Lipoic acid inhibits inflammatory bone resorption by suppressing prostaglandin E2 synthesis. J Immunol 2006 176:111–7.

Hamdy NA. Strontium ranelate improves bone microarchitecture in osteoporosis. Rheumatology. 2009 48S:9–13

Hammerman PS, Fox CJ, Birnbaum MJ, Thompson CB. Pim and Akt oncogenes are independent regulators of hematopoietic cell growth and survival. Blood. 2005 105:4477–83.

Hardy ML, Coulter I, Morton SC, Favreau J, Venuturupalli S, Chiappelli F, Rossi F, Orshansky G, Jungvig LK, Roth EA, Suttorp MJ, Shekelle P. S-adenosyl-L-methionine for treatment of depression, osteoarthritis, and liver disease. Evid Rep Technol Assess 2003 64:1–3.

Heinritz W, Hüffmeier U, Strenge S, Miterski B, Zweier C, Leinung S, et al. New mutations of EXT1 and EXT2 genes in German patients with Multiple Osteochondromas. Ann Hum Genet. 2009 73:283–91

Helfrich CD, Li YF, Sharp ND, Sales AE. Organizational readiness to change assessment (ORCA): development of an instrument based on the Promoting Action on Research in Health Services (PARiHS) framework. Implement Sci. 2009 4:38.

Herford AS, Boyne PJ, Rawson R, Williams RP. Bone morphogenetic protein-induced repair of the premaxillary cleft. J Oral Maxillofac Surg. 2007 65:2136–41.

Higgins JPT, Green S. Eds Cochrane Handbook for Systematic Reviews of Interventions. The Cochrane Collaboration 2008. Available from http://www.cochrane-handbook.org.

Hiraga T, Myoui A, Choi ME, Yoshikawa H, Yoneda T. Stimulation of cyclooxygenase-2 expression by bone-derived transforming growth factor-beta enhances bone metastases in breast cancer. Cancer Res 2006 66:2067–73.

Hofbauer LC, Khosla S, Dunstan C, Lacey D, Spelsberg T, Riggs B. Estrogen stimulates gene expression and protein production of osteoprotegerin in human osteoblastic cells. Endocrinology 1999 140:4367–70.

Hoffman SJ, Lavis JN, Bennett S. The Use of Research Evidence in Two International Organizations' Recommendations about Health Systems. Health Pol 2009 5:66–86.

Holtmeier W, Kabelitz D. γδ T cells link innate and adaptive immune responses. Chem Immunol Allergy. 2005 86:151–83.

Horwitz, EM. Building bone from blood vessels. Nature 2010, 16:1373–4.

Hu MI, Lu H, Gagel RF. Cancer therapies and bone health. Curr Rheumatol Rep. 2010 12:177–85.

Huang Y, Harrison J, Lorenzo J, Kream B. Parathyroid hormone induces interleukin-6 heterogeneous nuclear and messenger RNA expression in murine calvarial organ cultures. Bone 1998 23:327–32.

Huang B, Sun Y, Maciejewska I, Qin D, Peng T, McIntyre B, Wygant J, Butler WT, Qin C. Distribution of SIBLING proteins in the organic and inorganic phases of rat dentin and bone. Eur J Oral Sci. 2008 116:104–12.

Hughes DE, Dai A, Tiffee JC, Li HH, Mundy GR, Boyce BF. Estrogen promotes apoptosis of murine osteoclasts mediated by TGF-beta. Nat Med 1996 2:1132–6.

Hwang SY, Kim JY, Kim KW, Park MK, Moon Y, Kim WU, Kim HY. IL-17 induces production of IL-6 and IL-8 in rheumatoid arthritis synovial fibroblasts via NF-kappaB- and PI3-kinase/Akt-dependent pathways. Arthritis Res Ther. 2004 6:R120–8.

Ikeda R, Yoshida K, Inoue I. Identification of FAZF as a novel BMP2-induced transcription factor during osteoblastic differentiation. J Cell Biochem. 2007 101:147–54.

Indumati V, Patil VS. Biochemical Markers of Bone Remodelling in Osteoporosis. J Clinical Diagnostic Res 2010, 3:2089–97.

Isales CM, Zaidi M, Blair HC. ACTH is a novel regulator of bone mass. Ann NY Acad Sci. 2010 1192:110–6.

Jackson DG. Immunological functions of hyaluronan and its receptors in the lymphatics. Immunol Rev. 2009 230:216–31.

Jean-Pierre David. Advances in Immunology, 2007, 95: 149–165

Jennes I, Pedrini E, Zuntini M, Mordenti M, Balkassmi S, Asteggiano CG, Casey B, Bakker B, Sangiorgi L, Wuyts W. Multiple osteochondromas: mutation update and description of the multiple osteochondromas mutation database (MOdb). Hum Mutat. 2009 30:1620–7

Jiang J, Lv HS, Lin JH, Jiang DF, Chen ZK LTB4 can directly stimulate human osteoclast formation from PBMC independent of RANKL. Artif Cells Blood Substit Immobil Biotechnol 2005, 33:391–403.

Jilka RL, Hangoc G, Girasole G, Passeri G, Williams DC, Abrams JS, Boyce B, Broxmeyer H, Manolagas SC. Increased osteoclast development after estrogen loss: mediation by interleukin-6. Science 1992 257:88–91.

Jilka RL, Passeri G, Girasole G, Cooper S, Abrams J, Broxmeyer H, Manolagas SC. Estrogen loss upregulates hematopoiesis in the mouse: a mediating role of IL-6. Exp Hematol 1995 23:500–6.

Jilka RL, Weinstein RS, Bellido T, Parfitt AM, Manolagas SC. Osteoblast programmed cell death (apoptosis): modulation by growth factors and cytokine. J Bone Miner Res 1998 13:793–802.

John V, Hock JM, Short LL, Glasebrook AL, Galvin RJS. A role for CD8+ T lymphocytes in osteoclast differentiation in vitro. Endocrinology 1996 37:2457–63.

Jüni P, Egger M. PRISMAtic reporting of systematic reviews and meta-analyses. Lancet. 2009 374:1221–3.

Kearney F, Moore AR, Donegan CF, Lambert J. The ageing of HIV: implications for geriatric medicine. Age Ageing. 2010 Sep;39(5):536–41.

Khosla S. The OPG/RANKL/RANK system. Endocrinology. 2001 142:5050–5

Khosla S, Melton LJ, Riggs B. Estrogen and the male skeleton. J Clin Endocrinol Metab 2002 87:1443–50.

Kiechl S, Werner P, Knoflach M, Furtner M, Willeit J, Schett G. The osteoprotegerin/RANK/RANKL system: a bone key to vascular disease. Expert Rev Cardiovasc Ther. 2006 4:801–11.

Kim KW, Cho ML, Park MK, Yoon CH, Park SH, Lee SH, Kim HY. Increased interleukin-17 production via a phosphoinositide 3-kinase/Akt and nuclear factor kappaB-dependent pathway in patients with rheumatoid arthritis. Arthritis Res Ther. 2005 7:R139–48.

Kitazawa S, Kajimoto K, Kondo T, Kitazawa R. Vitamin D3 supports osteoclastogenesis via functional vitamin D response element of human RANKL gene promotor. J Cell Biochem 2003 89:771–7.

Kitson AL, Rycroft-Malone J, Harvey G, McCormack B, Seers K, Titchen A. Evaluating the successful implementation of evidence into practice using the PARiHS framework: theoretical and practical challenges. Implement Sci. 2008 3:1.

Kobayashi N, Kadono Y, Naito A, Matsumoto K, Yamamoto T, Tanaka S, Inoue J. Segregation of TRAF6-mediated signaling pathways clarifies its role in osteoclastogenesis. EMBO J 2001 20:1271–80.

Kogianni G, Mann V, Noble BS: Apoptotic osteocytes induce the production of pro-inflammatory osteoclastogenic cytokines. 1st Int Conf Osteoimmunology, 2006, A37.

Kolahi J, Bang H, Park J, Desbiens NA. CONSORT 2010 and Controversies Regarding Assessment of Blindness in RCTs. Dental Hypotheses 2010 1:99–105.

Kolls JK, Linden A. Interleukin-17 family members and inflammation. Immunity 2004 21:467–76.

Komarova SV, Smith RJ, Dixon SJ, Sims SM, Wahl LM. Mathematical model preedicts a critical role for osteoclast autocrine regulation in the cotrol of bone remodeling. Bone 2003 33:206–15

Komori T, Yagi H, Nomura S, Yamaguchi A, Sasaki K, Deguchi K, Shimizu Y, Bronson RT, Gao YH, Inada M, Sato M, Okamoto R, Kitamura Y, Yoshiki S, Kishimoto S. Targeted disruption of Cbfa1 results in a complete lack of bone formation owing to maturational arrest of osteoblasts. Cell 1997 89:755–64.

Kotake S, Udagawa N, Hakoda M, Mogi M, Yano K, Tsuda E, Takahashi K, Furuya T, Ishiyama S, Kim KJ, Saito S, Nishikawa T, Takahashi N, Togari A, Tomatsu T, Suda T, Kamatani N. Activated human T cells directly induce osteoclastogenesis from human monocytes: possible role of T cells in bone destruction in rheumatoid arthritis patients. Arthritis Rheum. 2001 44:1003–12.

Kronenberg H, Bringhurst F, Nussbaum S, Jüppner H, Abou-Samra A, Segre G, Potts J: Parathyroid hormone: biosynthesis, secretion, chemistry, and action; in Mundy G, Martin J (eds): Handbook of Experimental Pharmacology: Physiology and Pharmacology of Bone. Heidelberg, Springer, 1993, pp 185–201.

Kung J, Chiappelli F, Cajulis OS, Avezova R, Kossan G, Chew L, Maida CA. From Systematic Reviews to Clinical Recommendations for Evidence-Based Health Care: Validation of Revised Assessment of Multiple Systematic Reviews (R-AMSTAR) for Grading of Clinical Relevance. The Open Dentistry Journal 2010, 4:84–91.

Kwak HB, Ha H, Kim HN, Lee JH, Kim HS, Lee S, Kim HM, Kim JY, Kim HH, Song YW, Lee ZH. Reciprocal cross-talk between RANKL and interferon-gamma-inducible protein 10 is responsible for bone-erosive experimental arthritis. Arthritis Rheum. 2008 58:1332–42.

Lachin JM. Statistical Considerations in the Intent-to-Treat Principle. Controlled Clinical Trials 2000, 21:167–18.

LeBlanc A, Kenny DA, O'Connor AM, Légaré F. Decisional conflict in patients and their physicians: a dyadic approach to shared decision making. Med Decis Making. 2009 29:61–8.

Leder B, LeBlanc K, Schoenfeld D, Eastell R, Finkelstein J. Differential effects of androgens and estrogens on bone turnover in normal men. J Clin Endocrinol Metab 2003 88:204–10.

Lee MD. Three case studies in the Bayesian analysis of cognitive models. Psychon Bull Rev. 2008 15:1–15.

Lee NJ, Herzog H. NPY regulation of bone remodelling. Neuropeptides. 2009 43:457–63.

Lee NK, Sowa H, Hinoi E, Ferron M, Ahn JD, Confavreux C, Dacquin R, Mee PJ, McKee MD, Jung DY, Zhang Z, Kim JK, Mauvais-Jarvis F, Ducy P, Karsenty G. Endocrine regulation of energy metabolism by the skeleton. Cell 2007, 130: 456–69

Lee SH, Kim TS, Choi Y, Lorenzo J. Osteoimmunology: cytokines and the skeletal system. BMB Rep. 2008 41:495–510.

Légaré F, Moher D, Elwyn G, LeBlanc A, Gravel K. Instruments to assess the perception of physicians in the decision-making process of specific clinical encounters: a systematic review. BMC Med Inform Decis Mak. 2007 7:30.

Leibbrandt A, Penninger JM. RANK/RANKL: regulators of immune responses and bone physiology. Ann N Y Acad Sci. 2008 1143:123–50.

Levy AG, Hershey JC. Value-induced bias in medical decision making. Med Decis Making. 2008 28:269–76.

Li L, Pettit AR, Gregory LS, Forwood MR. Regulation of bone biology by prostaglandin endoperoxide H synthases (PGHS): a rose by any other name. Cytokine Growth Factor Rev. 2006 17:203–16

Li B, Wang P, Li H. The association between TNF-alpha promoter polymorphisms and ankylosing spondylitis: a meta-analysis. Clin Rheumatol. 2010 29:983–90.

Liberati A, Altman DG, Tetzlaff J, Mulrow C, Gøtzsche PC, Ioannidis JP, Clarke M, Devereaux PJ, Kleijnen J, Moher D. The PRISMA statement for reporting systematic reviews and meta-analyses of studies that evaluate health care interventions: explanation and elaboration. PLoS Med. 2009 6:e1000100.

Lin D, Rieder MJ. Interventions for the treatment of decreased bone mineral density associated with HIV infection. Cochrane Database Syst Rev. 2007 18:CD005645.

Lin KW, Slawson DC. Identifying and using good practice guidelines. Am Fam Physician. 2009 80:67–70.

Little J, Higgins JP, Ioannidis JP, Moher D, Gagnon F, von Elm E, Khoury MJ, Cohen B, Davey-Smith G, Grimshaw J, Scheet P, Gwinn M, Williamson RE, Zou GY, Hutchings K, Johnson CY, Tait V, Wiens M, Golding J, van Duijn C, McLaughlin J, Paterson A, Wells G, Fortier I, Freedman M, Zecevic M, King R, Infante-Rivard C, Stewart A, Birkett N. STrengthening the REporting of Genetic Association studies (STREGA): an extension of the STROBE Statement. Ann Intern Med. 2009 150:206–15

Lorenzo J, Choi Y. Osteoimmunology. Immunol. Rev. 2005 208:5–6.

Lorenzo J, Horowitz M, Choi Y. Osteoimmunology: interactions of the bone and immune system. Endocr Rev. 2008 29:403–40.

Lu T, Achari Y, Sciore P, Hart DA. Estrogen receptor alpha regulates matrix metalloproteinase-13 promoter activity primarily through the AP-1 transcriptional regulatory site. Biochim Biophys Acta. 2006 1762:719–31.

Lyles KW, Colón-Emeric CS, Magaziner JS, Adachi JD, Pieper CF, Mautalen C, Hyldstrup L, Recknor C, Nordsletten L, Moore KA, Lavecchia C, Zhang J, Mesenbrink P, Hodgson PK, Abrams K, Orloff JJ, Horowitz Z, Eriksen EF, Boonen S; HORIZON Recurrent Fracture Trial. Zoledronic acid and clinical fractures and mortality after hip fracture. N Engl J Med. 2007 357:1799–809.

Macdonald-Jankowski DS. Focal cemento-osseous dysplasia: a systematic review. Dentomaxillofac Radiol. 2008 37:350–60.

MacDonald-Jankowski DS. Ossifying fibroma: a systematic review. Dentomaxillofac Radiol. 2009 38:495–513.

MacPherson H, Altman DG, Hammerschlag R, Youping L, Taixiang W, White A, Moher D. STRICTA Revision Group. Revised STandards for Reporting Interventions in Clinical Trials of Acupuncture (STRICTA): extending the CONSORT statement. PLoS Med. 2010 7:e1000261.

Manheimer E, Cheng K, Linde K, et al. Acupuncture for peripheral joint osteoarthritis. Cochrane Database Syst Rev 2010 (1):CD001977.

Marhaba R, Nazarenko I, Knöfler D, Reich E, Voronov E, Vitacolonna M, Hildebrand D, Elter E, Apte RN, Zöller M. Opposing effects of fibrosarcoma cell-derived IL-1 alpha and IL-1 beta on immune response induction. Int J Cancer 2008 123:134–45.

Marina N, Gorlick R. Immune approaches to treating osteosarcoma. Cancer Biol Ther. 2009 8:981–3.

Marriott I, Rati D, McCall S, Tranguch S. Induction of Nod1 and Nod2 intracellular pattern recognition receptors in murine osteoblasts following bacterial challenge. Infect Immun 2005 73:2967–73.

Martin TJ, Seeman E. Bone remodelling: its local regulation and the emergence of bone fragility. Best Pract Res Clin Endocrinol Metab. 2008 22:701–22.

Matsuo K, Irie N. Osteoclast-osteoblast communication. Arch Biochem Biophys. 2008 473:201–9.

McClung MR, Lewiecki EM, Cohen SB, Bolognese MA, Woodson GC, Moffett AH, Peacock M, Miller PD, Lederman SN, Chesnut CH, Lain D, Kivitz AJ, Holloway DL, Zhang C, Peterson MC, Bekker PJ. AMG 162 Bone Loss Study Group: Denosumab in postmenopausal women with low bone mineral density. N Engl J Med 2006 354:860–3.

McLeod C, Bagust A, Boland A, Dagenais P, Dickson R, Dundar Y, Hill RA, Jones A, Mujica Mota R, Walley T. Adalimumab, etanercept and infliximab for the treatment of ankylosing spondylitis: a systematic review and economic evaluation. Health Technol Assess. 2007 11:1–158, iii-iv.

Meckley LM, Greenberg D, Cohen JT, Neumann PJ. The adoption of cost-effectiveness acceptability curves in cost-utility analyses. Med Decis Making. 2010 30:314–9.

Medici D, Shore EM, Lounev VY, Kaplan FS, Kalluri R, Olsen BR. Conversion of vascular endothelial cells into multipotent stem-like cells. Nature Medicine 2010, 16:1400–6.

Menezes GB, McAvoy EF, Kubes P. Hyaluronan, platelets, and monocytes: a novel pro-inflammatory triad. Am J Pathol. 2009 174:1993–5.

Merry K, Dodds R, Littlewood A, Gowen M. Expression of Osteopontin mRNA by osteoclasts and osteoblasts in modelling adult human bone. J Cell Sci 1993, 104:1013–20

Migliore A, Giovannangeli F, Bizzi E, Massafra U, Alimonti A, Laganà B, Diamanti Picchianti A, Germano V, Granata M, Piscitelli P. Viscosupplementation in the management of ankle osteoarthritis: a review. Arch Orthop Trauma Surg. 2011 131:139–47.

Mobley LR, Hoerger TJ, Wittenborn JS, Galuska DA, Rao JK. Cost-effectiveness of osteoporosis screening and treatment with hormone replacement therapy, raloxifene, or alendronate. Med Decis Making. 2006 26:194–206.

Moher D, Cook DJ, Eastwood S, Olkin I, Rennie D, Stroup DF. Improving the quality of reports of meta-analyses of randomised controlled trials: the QUOROM statement. Quality of Reporting of Meta-analyses. Lancet. 1999 354:1896–900.

Moher D, Liberati A, Tetzlaff J, Altman DG. PRISMA Group. Preferred reporting items for systematic reviews and meta-analyses: the PRISMA statement. PLoS Med. 2009 6:e1000097.

Mondy K, Tebas P. Emerging bone problems in patients infected with human immunodeficiency virus. Clin Infect Dis. 2003 36:S101–5.

Montori VM, Guyatt GH. Progress in evidence-based medicine. JAMA 2008 300:1814–6.

Morimoto J, Kon S, Matsui Y, Uede T. Osteopontin; as a target molecule for the treatment of inflammatory diseases. Curr Drug Targets. 2010 11:494–505.

Moss, ML. The non-existent hinge axis, Am. Inst. Oral Biol. 1972, 59–66

Murthy RG, Reddy BY, Ruggiero JE, Rameshwar P. Tachykinins and hematopoietic stem cell functions: implications in clinical disorders and tissue regeneration. Front Biosci. 2007, 12:4779–87.

Nakashima T, Takayanagi H. Osteoimmunology: crosstalk between the immune and bone systems. J Clin Immunol. 2009 29:555–67.

Nandi P, Kingsley GH, Scott DL. Disease-modifying antirheumatic drugs other than methotrexate in rheumatoid arthritis and seronegative arthritis. Current Opinion in Rheumatology 2008 20:251–6.

Noble BS, Peet N, Stevens HY, Brabbs A, Mosley JR, Reilly GC, Reeve J, Skerry TM, Lanyon LE. Mechanical loading: biphasic osteocyte survival and targeting of osteoclasts for bone destruction in rat cortical bone. Am J Physiol 2003 284:C934–C44.

Northrop JP, Ho SN, Chen L, Thomas DJ, Timmerman LA, Nolan GP, Admon A, Crabtree GR. NF-AT components define a family of transcription factors targeted in T-cell activation. Nature 1994 369: 497–502.

Oakley JE, O'Hagan A. Probabilistic sensitivity analysis of complex models: a Bayesian approach. JR Statist Soc B 2004 66:751–69.

Ofotokun I, Weitzmann MN. HIV-1 infection and antiretroviral therapies: risk factors for osteoporosis and bone fracture. Curr Opin Endocrinol Diabetes Obes. 2010 17:523–9.

Ogilvie D, Craig P, Griffin S, Macintyre S, Wareham NJ. A translational framework for public health research. BMC Public Health. 2009 9:116.

Oh KI, Kim BK, Ban YL, Choi EY, Jung KC, Lee IS, Park SH. CD99 activates T cells via a costimulatory function that promotes raft association of TCR complex and tyrosine phosphorylation of TCR zeta. Exp Mol Med 2007 39:176–84.

Onyia J, Libermann T, Bidwell J, Arnold D, Tu Y, McClelland P, et al: Parathyroid hormone (1–34)-mediated interleukin-6 induction. J Cell Biochem 1997 67:265–74.

Otto F, Thornell AP, Crompton T, Denzel A, Gilmour KC, Rosewell LR, Stamp GWH, Beddington RSP, Mundlos S, Olsen BP, Selby PB, Owen MJ. Cbfa1, a candidate gene for cleidocranial dysplasia syndrome, is essential for osteoblast differentiation and bone development. Cell 1997 89:765–77.

Oursler MJ, Cortese C, Keeting PE, Anderson MA, Bonde SK, Riggs BL, Spelsberg TC. Modulation of transforming growth factor-beta production in normal human osteblast-like cells by 17β-estrodiol and parathyroid hormone. Endocrinology 1991 129:3313–20.

Pacifici R, Carano A, Santoro SA, Rifas L, Jeffrey JJ, Malone JD, McCracken R, Avioli LV. Bone matrix constituents stimulate interleukin-1 release from human blood mononuclear cells. J Clin Invest 1991 87:221–228.

Panda DK, Miao D, Bolivar I, Li J, Huo R, Hendy GN, Glotzman D. Inactivation of the 25-hydroxylase and vitamin D receptor demonstrates independent and interdependent effects of calcium and vitamin D on skeletal and mineral homeostasis. J Biol Chem 2004 179:16754–66.

Park JB. Use of bone morphogenetic proteins in sinus augmentation procedure. J Craniofac Surg. 2009 20:1501–3.

Park JB. Use of cell-based approaches in maxillary sinus augmentation procedures. J Craniofac Surg. 2010 21:557–60.

Pazianas M, Miller P, Blumentals WA, Bernal M, Kothawala P. A review of the literature on osteonecrosis of the jaw in patients with osteoporosis treated with oral bisphosphonates: prevalence, risk factors, and clinical characteristics. Clin Ther. 2007 29:1548–58.

Pérez-Sayáns M, Somoza-Martín JM, Barros-Angueira F, Rey JM, García-García A. RANK/RANKL/OPG role in distraction osteogenesis. Oral Surg Oral Med Oral Pathol Oral Radiol Endod. 2010 109:679–86.

Philip BK, Childress PJ, Robling AG, Heller A, Nawroth PP, Bierhaus A, Bidwell JP. RAGE supports parathyroid hormone-induced gains in femoral trabecular bone. Am J Physiol Endocrinol Metab. 2010 298:E714–25.

Pilbeam CC, Harrison JR, Raisz LG: Prostaglandins and bone metabolism; in Bilezikian JP, Raisz LG, Rodan GA (eds): Principles of Bone Biology. San Diego, Academic, 2002, pp 979–994.

Plows D, Kontogeorgos G, Kollias G. Mice lacking mature T and B lymphocytes develop arthritic lesions after immunization with type II collagen. J Immunol 1999 162:1018–23.

Pollock J, Blaha M, Gornik S, Stevenson S, Greenfield E. In vivo demonstration that parathyroid hormone and parathyroid hormone-related protein stimulate expression by osteoblasts of interleukin-6 and leukemia inhibitory factor. J Bone Miner Res 1996 11:754–9.

Potts J, Jüppner H. Parathyroid hormone and parathyroid hormone-related peptide in calcium homeostasis, bone metabolism, and bone development: the proteins, their genes, and receptors; in L. Avioli, S. Krane, Eds: Metabolic Bone Disease. New York, Academic, 1997, pp 51–94.

Prochaska JO, Decision Making in the Transtheoretical Model of Behavior Change. Med Decis Making 2008 28:845–9.

Prolo P, Chiappelli F, Bernard G, Fiala M, Ibarra A, Sartori ML, Dovio A, Angeli A. Neuroimmune surveillance of osteosarcoma: emerging hypotheses. Journal of Dental Research, 2003 82:417–21.

Quinn JMW & Saleh H. (2009). Modulation of osteoclast function in bone by the immune system. Molecular and Cellular Endocrinology 310, 40–51

Rainio EM, Sandholm J, Koskinen PJ. Cutting edge: Transcriptional activity of NFATc1 is enhanced by the Pim-1 kinase. J. Immunol. 2002 168:1524–7.

Rauner M, Sipos W, Pietschmann P. Osteoimmunology. Int Arch Allergy Immunol. 2007 143:31–48.

Reid P, Holen I. Pathophysiological roles of osteoprotegerin (OPG). Eur J Cell Biol. 2009 88:1–17

Reyna VF, A Theory of Medical Decision Making and Health: Fuzzy Trace Theory. Med Decis Making 2008 28:850–65

Riggs BL, Hartmann LC. Selective estrogen-receptor modulators: mechanisms of action and application to clinical practice. N Engl J Med 2003 348:618–29.

Ripamonti U, Herbst NN, Ramoshebi LN. Bone morphogenetic proteins in craniofacial and periodontal tissue engineering: experimental studies in the non-human primate Papio ursinus. Cytokine Growth Factor Rev. 2005 16:357–68.

Robey PG, Bianco P. The use of adult stem cells in rebuilding the human face. J Am Dent Assoc. 2006 137:961–72.

Rodríguez-Vázquez JF, Meyrida-Velasco JR, Meyrida-Velasco JA, Jimeynez-Collado J. Anatomical considerations on the discomalleolar ligament. J Anat. 1998 192: 617–21.

Roelofs AJ, Hulley PA, Meijer A, Ebetino FH, Russell RG, Shipman CM. Selective inhibition of Rab prenylation by a phosphonocarboxylate analogue of risedronate induces apoptosis, but not S-phase arrest, in human myeloma cells. Int J Cancer. 2006 119:1254–61.

Romagnani S, Maggi E, Liotta F, Cosmi L, Annunziato F. Properties and origin of human Th17 cells. Mol Immunol. 2009 47:3–7.

Rowicki T, Zakrzewska J. A study of the discomalleolar ligament in the adult human. Folia Morphol. (Warsz). 2006 65:121–5

Ruof J, Stucki G. Comparison of the Dougados Functional Index and the Bath Ankylosing Spondylitis Functional Index. A literature review. J Rheumatol. 1999 26:955–60.

Samson DJ, Grant MD, Ratko TA, Bonnell CJ, Ziegler KM, Aronson N. Treatment of Primary and Secondary Osteoarthritis of the Knee. Evidence Report/Technology Assessment No. 157 (Prepared by Blue Cross and Blue Shield Association Technology Evaluation Center Evidence-based Practice Center under Contract No. 290-02-0026). AHRQ Pub. 07-E012. Rockville: AHRQ. September 2007.

Schoppet M, Henser S, Ruppert V, Stübig T, Al-Fakhri N, Maisch B, Hofbauer LC. Osteoprotegerin expression in dendritic cells increases with maturation and is NF-kappaB-dependent. J Cell Biochem. 2007 100:1430–9.

Schulz KF, Altman DG, Moher D, CONSORT Grp. CONSORT 2010 statement: updated guidelines for reporting parallel group randomised trials. PLoS. Med. 2010 7:e1000251.

Schünemann HJ, Oxman AD, Brozek J, Glasziou P, Bossuyt P, Chang S, Muti P, Jaeschke R, Guyatt GH. GRADE: assessing the quality of evidence for diagnostic recommendations. Evid Based Med. 2008 13:162–3.

Scuffham PA, Whitty JA, Mitchell A, Viney R. The use of QALY weights for QALY calculations: a review of industry submissions requesting listing on the Australian Pharmaceutical Benefits Scheme 2002–4. Pharmacoeconomics. 2008 26:297–310.

Sereti I, Rodger AJ, French MA. Biomarkers in immune reconstitution inflammatory syndrome: signals from pathogenesis. Curr Opin HIV AIDS. 2010 5:504–10.

Shea BJ, Grimshaw JM, Wells GA, Boers M, Andersson N, Hamel C, Porter AC, Tugwell P, Moher D, Bouter LM. Development of AMSTAR: a measurement tool to assess the methodological quality of systematic reviews. BMC Med Res Methodol. 2007 7:10

Shea BJ, Hamel C, Wells GA, Bouter LM, Kristjansson E, Grimshaw J, Henry DA, Boers M. AMSTAR is a reliable and valid measurement tool to assess the methodological quality of systematic reviews. J Clin Epidemiol. 2009 62:1013–20.

Shi H, Lyons-Weiler J. Clinical decision modeling system. BMC Med Inform Decis Mak. 2007 7:23.

Shinoda K, Sugiyama E, Taki H, Harada S, Mino T, Maruyama M, Kobayashi M. Resting T cells negatively regulate osteoclast generation from peripheral blood monocytes. Bone 2003 33:711–20.

Shoji M, Tanabe N, Mitsui N, Tanaka H, Suzuki N, Takeichi O, Sugaya A, Maeno M: Lipopolysaccharide stimulates the production of prostaglandin E(2) and the receptor Ep4 in osteoblasts. Life Sci 2006 78:2012–8.

Simera I, Moher D, Hoey J, Schulz KF, Altman DG. The EQUATOR Network and reporting guidelines: Helping to achieve high standards in reporting health research studies. Maturitas. 2009 63:4–6.

Simera I, Moher D, Hirst A, Hoey J, Schulz KF, Altman DG. Transparent and accurate reporting increases reliability, utility, and impact of your research: reporting guidelines and the EQUATOR Network. BMC Med. 2010 8:24.

Simon D, Loh A, Härter M. Measuring (shared) decision-making – a review of psychometric instruments. Z Arztl Fortbild Qualitatssich. 2007 101:259–67.

Simone MJ, Appelbaum J. HIV in older adults. Geriatrics. 2008 63:6–12.

Simonet WS, Lacey DL, Dunstan CR, Kelley M, Chang MS, Luthy R, Nguyen HQ, Wooden S, Bennett L, Boone T, Shimamoto G, DeRose M, Elliott R, Colombero A, Tan HL, Trail G, Sullivan J, Davey E, Bucay N, Renshaw-Gregg L, Hughes TM, Hill D, Pattison W, Campell P, Boyle WJ. Osteoprotegerin: a novel secreted protein involved in regulation of bone density. Cell 1997 89:309–19.

Sousa DM, Herzog H, Lamghari M. NPY signalling pathway in bone homeostasis: Y1 receptor as a potential drug target. Curr Drug Targets. 2009 10:9–19.

Spring B. Health Decision Making: Lynchpin of Evidence-Based Practice. Med Decis Making 2008 28:866–74.

Staal SP et al. Isolation of transforming murine leukemia viruses from mice with a high incidence of spontaneous lymphoma. PNAS 1977 74:3065–7.

Stanley KT, VanDort Ch, Motyl Ch, Endres J, Fox DA. Immunocompetent properties of human osteoblasts: interactions with T lymphocytes. J Bone Miner Res 2006 1:29–36.

Stephens EA, Das R, Madge S, Barter J, Johnson MA. Symptomatic osteoporosis in two young HIV-positive African women. AIDS. 1999 13:2605–6.

Stevenson M, Jones ML, De Nigris E, Brewer N, Davis S, Oakley J. A systematic review and economic evaluation of alendronate, etidronate, risedronate, raloxifene and teriparatide for the prevention and treatment of postmenopausal osteoporosis. Health Technol Assess. 2005 9:1–160.

Strum JM, Gartner LP, Hiatt JL. Cell biology and histology. Lippincott Williams & Wilkins, Hagerstwon. 2007.

Sun Y, Ma S, Zhou J, Yamoah A, Feng JQ, Hinton RJ, Qin C. Distribution of Small Integrin-binding LIgand, N-linked Glycoproteins (SIBLING) in the Articular Cartilage of the Rat Femoral Head. J Histochem Cytochem. 2010 58(11):1033–43.

Swallow DM, Povey S, Parkar M, Andrews PW, Harris H, Pym B, Goodfellow P. Mapping of the gene coding for the human liver/bone/kidney isozyme of alkaline phosphatase to chromosome 1. Ann Hum Genet 1998 50:229–35.

Taichman RS, Emerson SG. The role of osteoblasts in the hematopoietic microenvironment. Stem Cells 1998, 16:7–15.

Taichman RS, Reilly MJ, Emerson SG. The hematopoietic microenvironment: osteoblasts and the hematopoietic microenvironment. Hematology 2000 4:421–6.

Takayanagi H. Osteoimmunology: shared mechanisms and crosstalk between the immune and bone systems. Nat. Rev. Immunol. 2007 7: 292–304.

Takayanagi H. Osteoimmunology and the effects of the immune system on bone. Nat Rev Rheumatol. 2009 5:667–76.

Takayanagi H. New immune connections in osteoclast formation. Ann NY Acad Sci. 2010 1192:117–23.

Takayanagi H, Ogasawara K, Hida S, Chiba T, Murata S, Sato K, Takaoka A, Yokochi T, Oda H, Tanaka K, Nakamura K, Taniguchi T. T-cell-mediated regulation of osteoclastogenesis by signaling cross-talk between RANKL and IFN-gamma. Nature 2000 408:600–5.

Takayanagi H, Sato K, Takaoka A, Taniguchi T. Interplay between interferon and other cytokine systems in bone metabolism. Immunol Rev 2005 208:181–93.

Taniguchi T, Dandrea AD. Molecular pathogenesis of fanconi anemia. Int J Hematol. 2002 75:123–8.

Tebas P, Powderly WG, Claxton S, Marin D, Tantisiriwat W, Teitelbaum SL, Yarasheski KE. Accelerated bone mineral loss in HIV-infected patients receiving potent antiretroviral therapy. AIDS. 2000 14:F63–7.

ten Dijke P, Ichijo H, Franzén P, Schulz P, Saras J, Toyoshima H, Heldin CH, Miyazono K. Activin receptor-like kinases: a novel subclass of cell-surface receptors with predicted serine/threonine kinase activity. Oncogene 1993, 8: 2879–87.

Terracciano L, Brozek J, Compalati E, Schünemann H. GRADE system: new paradigm. Curr Opin Allergy Clin Immunol. 2010 10:377–83

Thelle DS. STROBE and STREGA: Instruments for improving transparency and quality of reporting scientific results. Eur J Epidemiol. 2009 24:7–8.

Theoleyre S, Wittrant Y, Tat SK, Fortun Y, Redini F, Heymann D. The molecular triad OPG/RANK/RANKL: involvement in the orchestration of pathophysiological bone remodeling. Cytokine Growth Factor Rev 2004 15:457–75.

Thomas J, Doherty SM. HIV infection – a risk factor for osteoporosis. J Acquir Immune Defic Syndr. 2003 33:281–91.

Titler MG. Methods in translation science. Worldviews Evid Based Nurs. 2004 1:38–48.

Tomazic J, Ul K, Volcansek G, Gorensek S, Pfeifer M, Karner P, Prezelj J, Vidmar G, Vidmar L. Prevalence and risk factors for osteopenia/osteoporosis in an HIV-infected male population. Wien Klin Wochenschr. 2007 119:639–46.

Traianedes K, Dallas MR, Garrett IR, Mundy GR, Bonewald LF. 5-Lipoxygenase metabolites inhibit bone formation in vitro. Endocrinology 1998, 139:3178–84.

van der Heijde D. New directions in classification and outcome assessment in ankylosing spondylitis. Curr Rheumatol Rep. 2004 6:98–101.

von Elm E, Altman DG, Egger M, Pocock SJ, Gøtzsche PC, Vandenbroucke JP. STROBE Initiative. The Strengthening the Reporting of Observational Studies in Epidemiology (STROBE)statement: guidelines for reporting observational studies. Lancet. 2007 370:1453–7.

von Elm E, Moher D, Little J. STREGA collaboration. Reporting genetic association studies: the STREGA statement. Lancet. 2009 374:98–100.

Wang KX, Denhardt DT. Osteopontin: role in immune regulation and stress responses. Cytokine Growth Factor Rev. 2008 19:333–45.

Warden SJ, Bliziotes MM, Wiren KM, Eshleman AJ, Turner CH. Neural regulation of bone and the skeletal effects of serotonin (5-hydroxytryptamine). Mol Cell Endocrinol. 2005 242:1–9.

Watkins BA, Li Y, Lippman HE, Feng S. Modulatory effect of omega-3 polyunsaturated fatty acids on osteoblast function and bone metabolism. Prostaglandins Leukot Essent Fatty Acids. 2003 68:387–98.

Weber GF, Cantor H. The immunology of Eta-1/osteopontin. Cytokine Growth Factor Rev. 1996 7:241–8

Weiel JE, Lenhard JM. Bone mineral loss in HIV-positive patients receiving antiretroviral therapy. AIDS. 2000 14:2218–9.

Weiss T, Vitacolonna M, Zöller M. The efficacy of an IL-1alpha vaccine depends on IL-1RI availability and concomitant myeloid-derived suppressor cell reduction. J Immunother. 2009 32:552–64

Weitzmanm MN and Pacifici R. Ann N Y Acad Sci, 2007. 1116: 360–75

Weitzmann MN and Pacifici R. Immunological Reviews, 2005b, 208:5–280

Weitzmann MN and Pacifici R. Bone and the Immune System. Bone Disease of Organ Transplantation 2005a, 91–109

West SL, Gartlehner G, Mansfield AJ, Poole C, Tant E, Lenfestey N, Lux LJ, Amoozegar J, Morton SC, Carey TC, Viswanathan M, Lohr KN. Clinical Heterogeneity. In: AHRQ Report 10-EHC-070 EF: Comparative Effectiveness Review Methods. Rockville, MD. 2010.

Whitney SN, Holmes-Rovner M, Brody H, Schneider C, McCullough LB, Volk RJ, McGuire AL. Beyond shared decision making: an expanded typology of medical decisions. Med Decis Making. 2008 28:699–705.

Williams RO, Paleolog E, Feldmann M. Cytokine inhibitors in rheumatoid arthritis and other autoimmune diseases. Curr Opin Pharmacol 2007 7: 412–7.

Wolf G. Energy regulation by the skeleton. Nutr Rev. 2008 66:229–33.

Women's Health Initiative. Effects of conjugated equine estrogen in postmenopausal women with hysterectomy. JAMA 2004 291:1701.

Women's health Initiative. Risks and benefits of estrogen plus progestin in healthy postmenopausal women. JAMA 2002 288:321.

Wong IP, Baldock PA, Herzog H. Gastrointestinal peptides and bone health. Curr Opin Endocrinol Diabetes Obes. 2010, 17:44–50.

Wyzga N, Varghese S, Wikel S, Canalis E, Sylvester FA. Effects of activated T cells on osteoclastogenesis depend on how they are activated. Bone 2004 35:614–20.

Xing L. Regulation of osteoclast precursors in inflammatory bone loss. Current Opinions Investigational Drugs. 2009, 10:1195–203

Yamashita T, Nakahata T. Current knowledge on the pathophysiology of Fanconi anemia: from genes to phenotypes. Int J Hematol. 2001 74:33–41.

Yasuda H, Shima N, Nakagawa N, Yamaguchi K, Kinosaki M, Mochizuki S, Tomoyasu A, Yano K, Goto M, Murakami A, Tsuda E, Morinaga T, Higashio K, Udagawa N, Takahashi N, Suda T: Osteoclast differentiation factor is a ligand for osteoprotegerin/osteoclastogenesis-inhibitory factor and is identical to TRANCE/RANKL. Proc Natl Acad Sci USA 1998 95:3597–602.

Yaykasli KO, Oohashi T, Hirohata S, Hatipoglu OF, Inagawa K, Demircan K, Ninomiya Y. ADAMTS9 activation by interleukin 1 beta via NFATc1 in OUMS-27 chondrosarcoma cells and in human chondrocytes. Mol Cell Biochem. 2009 323:69–79.

Ye H, Arron JR, Lamothe B, Cirilli M, Kobayashi T, Shevde NK, Segal D, Dzivenu OK, Vologodskaia M, Yim M, Du K, Singh S, Pike JW, Darnay BG, Choi Y, Wu H. Distinct molecular mechanism for initiating TRAF6 signaling. Nature 2002 418:443–7.

Yu JJ, Gaffen SL. Interleukin-17: a novel inflammatory cytokine that bridges innate and adaptive immunity. Front Biosci. 2008 13:170–7.

Yuan F, Song L, Qian L, Hu JJ, Zhang Y. Assembling an orchestra: Fanconi anemia pathway of DNA repair. Front Biosci. 2010 15:1131–49.

Zhivotovsky B, Orrenius S. Cell death mechanisms: cross-talk and role in disease. Exp Cell Res. 2010 316:1374–83.

Zhu JD, Pan HO, Suzuki F, Takigawa M. Proto-oncogene expression in a human chondrosarcoma cell line: HCS-2/8. Jpn J Cancer Res. 1994 85:364–71.

Index

A

Acceptable sampling analysis, 71, 73, 74, 77, 97
ACTIS, 84
Activated protein 1 (AP-1), 14
Activin-like kinase-2 (ALK2), 8
Acupuncture, 78, 96
Agency for Healthcare Research and Quality (AHRQ), 1, 69, 70, 75, 80, 92, 99
AIDSTRIALS, 84
AKT, 13, 14, 22
Alkaline phosphatase, 4, 10, 36, 41
American Recovery and Reinvestment Act (ARRA), PL-111–5,
Anepiphyses, 7
Ankylosing spondylitis
  aka, 51
  Bekhterev's disease, 50
  Bekhterev syndrome, 50
  Marie-Strümpell disease, 50
Anoxia, 59
Antigen-presenting cells (APC's), 12
Anti-neutrophil cytoplasmic antibodies (ANCA), 51
Antiretroviral therapy (ART), 47, 48, 50, 83, 84
Appraisal of Guidelines Research and Evaluation-Europe (AGREE), 93
Assessment of multiple systematic reviews instrument (AMSTAR), 71, 77, 96, 97
ATPase V0 domain (Atp6V0d2), 11

B

Bath ankylosing spondylitis disease activity index (BASDAI), 53, 87
Bath ankylosing spondylitis functional index (BASFI), 53, 87
Bayesian analysis, 78
B cell, 5, 10, 22, 24, 43
Biblioteca Regionàl de Medicina (BIREME), 88
BIOSIS, 79
Biphosphonates, 41, 44–46, 59, 60, 81, 83, 85, 87, 88
Bone formation, 4, 5, 8, 9, 11, 16, 19, 24, 36, 40, 41, 45, 48, 57
Bone gamma-carboxyglutamic acid-containing protein (BGLAP), 10, 14
Bone marrow, 3, 5–7, 17, 23, 24, 36, 43, 57, 59, 60, 66, 67
Bone mass, 4, 9, 15, 40, 41, 44, 60
Bone metabolism, 5, 6, 8–10, 15–17, 19, 24, 101
Bone mineralization, 46, 47, 59
Bone morphogenetic protein-4 (BMP4), 8
Bone remodeling, 7, 8, 16, 40, 43
Bone resorption, 4, 8, 9, 11–16, 19, 20, 24, 28, 34, 40, 41, 43–46, 48, 58, 60, 64, 83, 98
Bucco-pharyngeal membrane, 27

C

Calcification, 4, 56, 57
Calcitonin, 4, 11, 12, 44, 46, 85
cAMP-dependent protein kinase A, 14
Canaliculi, 6
Cancellous bone, 59
CBFA1 deficient mice, 24
C-cells, 12
CCR6, 23
CD4, 16, 17, 20–23, 48–50, 52, 57
CD8, 16, 17, 51, 52, 57
CD28, 17, 48
CD44, 79

CD54 (ICAM-1), 18
CD99, 57
CD161, 23
CD166 (ALCAM), 18
Centers for Education and Research on Therapeutics (CERT's), 99
Chemokine (CXCL10), 48
Chondrosarcoma, 55, 56
CINAHL, 79
Clinically relevant complex systematic reviews (CRCSR's), 5, 76, 79, 99
Cochrane Controlled Trial Register, 79
Colony stimulating factor 1 receptor (CSF1R)
  aka, 14
  CD115, 14
  M-CFSR, 14
Common leukocyte antigen, CD45, 57
Compact bone, 6, 7
Comparative effectiveness analysis and research, 5, 72, 83, 100
Condyle, 30–35, 64–66, 100
  condylar process, 31, 63
Confidence interval (CI), 49, 80, 81, 83–85
Consensus of the best available evidence, 74
Consolidated standard for randomized trials, CONSORT (and 2010 revision: CONSORT-10), 96
Core binding factor 1 (Cbfa1), 8, 24, 67
Crohn's disease, 40, 42
Cyclo-oxygenase 2 (COX-2), 15, 16, 54

D
Dendritic cell, 5, 10, 12–14, 18, 23
Denosumab, 43, 46, 83
Diaphysis, 6
Disease-modifying antirheumatic drugs (DMARD's), 54, 98
Downstream factor osterix, Osx, 8
Dual action bone agents (DABAs), 46

E
Early T-lymphocyte activation (ETA-1)
  aka, 10
  SPP1, 10
Effectiveness, 1, 2, 5, 36, 69, 72, 74–76, 78, 80–83, 85–89, 91–93, 95, 97, 99–101
Efficacy, 2, 50, 70, 74, 79–81, 83, 85–87, 89, 98, 101
EMBASE, 72, 79, 84, 85, 88
Endosteal cells, 7
Enhancing the quality and transparency of health research (EQUATOR), 95
Enthesis, 50
Epigenetic events, 7
Estrogen, 5, 12, 13, 19, 42–45, 81, 83, 85
Evidence-based decision-making, 1, 74, 76, 100
Evidence-based practice centers (EBPC's), 99
Ewing's sarcoma, 57
Excursion, 34, 35
Exostosin, 55

F
Facial cranium
  aka, 28
  facial skeleton, 25
  rostral skeleton, 19
  rostro-cranium, 25
  splanchnocranium, 25, 27
  viscerocranium, 25
Fanconi anemia, 66–67
Fibrosarcoma, 56, 57
Fibrous symphysis, 31
Fixed model, 86
Fuzzy trace theory, 90, 91

G
$\gamma\delta$ T-cell receptors, 22
Glaserian fissure, 33
Grading of recommendations, assessment, development and evaluation (GRADE), 76, 77, 93–95
Gray literature, 71, 72, 80, 82, 84–86, 97
GTPase, 60

H
Hardy-Weinberg principle, 86
Harvesian system, 6
Hematopoiesis, 3, 10, 16
Her-2, 58
Highly active antiretroviral therapy (HAART), 47, 48, 50, 85
HIV/AIDS, 47–50, 83–85
Homogeneity *vs.* heterogeneity, 71, 73
Hormone replacement therapy (HRT), 5, 81, 82
Hyaluronic acid, 78–80
Hyoid bone, 25, 27, 32
Hypercalcemia, 10
Hypophosphatasia, 43

I

IL-3, 18
IL-4, 17, 18, 20
IL-6, 5, 13, 18–22, 48
IL-7, 18, 49
IL-10, 18, 20
IL-11, 13, 21
IL-12, 18, 19, 23, 40
IL-13, 18
IL-17, 18, 20–23, 34, 40
IL-1β, 19–21, 23, 56, 57
IL-17R, 22, 23
Immovable synarthrodial joints, 25
Immune-mediated bone degenerative diseases, 11
Immune reconstitution inflammatory syndrome (IRIS), 48, 49
Incidence, 5, 53, 58, 66, 81–83
INF-μ, 13
Interstitial lamellae, 6
Intramembranous ossification, 26

K

Knock-out mice, 14

L

Lacuna, 6, 11
Leptin, 15
Leukotrienes, 15, 16
Level of the evidence, 73, 74, 76, 80, 93, 95–97
LILACS, 84, 88
Linear regression, 49
Lipoxygenase, 15
5-Lipoxygenase-activating protein, FLAP, 15
Logic model for decision-making, 74
Logistic polytomic regression, 49
Lyme disease, 40
Lymphatic, 58, 79

M

Macrophage-colony-stimulating factor (M-CSF), 14, 19, 20, 23
Major histocompatibility complex (MHC), 18, 51, 52
MALDI-TOF, 47
Mandible, 26, 30–36, 57, 59, 61–65, 88, 100
Markov model, 82, 90
Matrix metalloproteinases (MMP), 11, 41, 46
Maxilla, 28–30, 36, 37, 57, 61, 63, 100
Maxillary artery, 26, 34, 61–63
Meckel's cartilage, 26, 30
Medline, 72, 79, 82–85, 88
Medullary bone
  aka,
  cancellous, 7
  spongy, 7
  trabecular, 7
MeSH terms, 72
Meta-analysis, 71, 73, 74, 77, 78, 81, 84–86, 97, 98
Metaphyseal region, 57
Microvillus structure, 11
Muscles of mastication, 26, 35, 61–63
Musculature of facial expression, 27
Myeloid lineage, 5

N

National Institute for Health and Clinical Excellence (NICE), 86
Neurocranium, 25
Neuropeptides, 5, 6
Non-steroidal anti-inflammatory drugs (NSAID's), 53, 54, 78, 86
Nuclear factor of activated T cells (NFAT), 14, 56
Nucleotide-binding oligomerization domain proteins 1 & 2 (NOD1 & NOD2), 18

O

Odds ratio, 49, 84, 85
Organizational Readiness to Change Assessment (ORCA), 92
Osteoarthritis, 1, 30, 31, 34, 39–40, 51, 64, 78–80
Osteoblasts, 4–13, 15–19, 23, 24, 36, 43, 44, 46, 59
Osteocartilaginous exostosis, 55
Osteocavitation, 60
Osteochondromas, 55–56
Osteoclastogenesis, 4, 9, 13, 14, 16–20, 34, 37, 40
Osteoclasts, 4–6, 8, 9, 11–15, 17–20, 23, 24, 41, 43–46, 59, 60, 67, 69
Osteocytes, 4, 6–8, 10, 11, 43, 59
Osteoid, 4, 10, 57, 59
Osteon, 6, 56
Osteonecrosis, 2, 40, 45, 47, 58–61, 87–88, 100
Osteonecrosis of the jaw, 2, 45, 59–61, 87–88

Osteopenia, 47, 49, 50, 54, 84, 85
Osteopontin (OPN)
aka,
BNSP, 10
BSP-1, 10
Osteoporosis, 5, 9, 12, 34, 40–50, 54, 60, 64, 80–85, 87, 100
Osteoprotegerin (OPG)
aka, 4
OCIF, 4, 12
osteoclastogenesis inhibitory factor, 4, 9, 13, 17, 19, 37
Osteosarcoma, 57, 58
Otic ganglion, 62, 65
Overview Quality Assessment Questionnaire (OQAQ), 96

P

Palatine, 25, 28–29, 62
Parathyroid hormone (PTH), 4, 11–13, 18, 19, 21, 44, 45, 81
Periosteum, 7, 25, 56
Perthes' disease, 40
Pharingo-laryngeal structures, 27
Pharyngeal arches, 26
Phosphatidylinositol–3 kinase, (PI3K), 14, 22
Phosphoinositide dependent kinase 1 (PDPK1), 13
PICO, 70, 72–74, 76, 84, 93, 97
PICOTS, 70, 71
Pim-1 kinase, 56
Platelet-derived growth factor (PDGF), 14
Preferred reporting items for systematic reviews and meta-analyses (PRISMA), 98
Prevalence, 28, 47, 49, 50, 83, 88
Programmed cell death, apoptosis, 11, 58
Programmed death, PD-1, 48
Pro-inflammatory and pro-resorptive cytokines, 5
Promoting Action on Research Implementation in Health Services (PARiHS), 92, 93
Protease inhibitors (PI), 14, 48, 50, 83, 84
Proteomic signature, 10, 13, 36, 76, 91, 101
Protrusion, 28, 35, 62, 63

Q

Quality adjusted life years (QALY's), 81, 82, 87
Quality assessment of studies of diagnostic accuracy included in systematic reviews (QUADAS), 96
Quality of reporting of meta-analyses (QUOROM), 98
Quality of the evidence, 71, 73, 74, 77, 80, 93–97

R

Randomized controlled trials (RTC's), 5, 79, 81, 84, 86, 95
Random models, 81, 86
RANKL(1-3) gene, 13
Receptor activator for nuclear factor κB ligand (RANKL)
aka, 4
ODF, 4, 13
OPGL, 4, 12
TRANCE, 4, 12
Receptor activator of nuclear factor κB, (RANK), 4, 12, 14, 17, 19–21, 23, 37, 40, 43, 44, 46, 83
Regulatory T cells (Tregs), 49, 57
Reichert's cartilage, 27
Reliability, 52, 72, 73, 79, 92
Research synthesis, 2, 5, 69–84, 86, 87, 93–101
Retinoic acid-related orphan receptor (ROR)γτ, 23
Retrusion, 35
Rheumatoid arthritis, 13, 16, 18, 22, 34, 40, 54, 77–80, 83, 98
Runt-related transcription factor 2 (runx2), 8, 67

S

Sacroiliac joints, 52, 53
S-adenosyl methionine, 78
SELDI-TOF, 47
Selective estrogen receptor modulator (SERM), 5, 45, 81, 83
Serine/threonine-specific protein kinase family, 13
Sharpey's fibers, 25, 36
Signal transducers and activators of transcription-1 (Stat1), 18
Small integrin-binding ligand, N-linked glycoproteins (SIBLING), 10
Standards for reporting interventions in clinical trials of acupuncture (STRICTA), 96
Staphylococcal enterotoxin A, 17
Stem cell niches, 24, 35
Stomatological ossification, 25–31, 33
Stomodeum, 27
Strengthening the reporting of genetic association studies (STREGA), 96

Strengthening the reporting of observational studies in epidemiology (STROBE), 96
Strength of Recommendation Taxonomy (SORT), 94
Stromal cells, 12, 17–19, 22, 43
Strontium ranelate, 12, 46, 81, 83
Substance P, 6
Suppressor of cytokine signaling 3 (SOCS3), 18
Synovial fibroblasts, 18, 22
Systematic review, 5, 71, 72, 74–88, 94–99

T
Tachykinins, 6
T cell, 5, 10, 13, 16, 18–24, 40, 42, 43, 48, 49, 51, 52, 56, 57, 60
T cell receptor (TcR), 22, 52, 57, 60
Temporomandibular joint, 30, 32–36, 61, 66, 99, 100
Temporomandibular joint disorder (TMD), 33, 61, 65
Temporomandibular joint ligaments, 35, 64
Teriparatide, 45, 81, 85
TH1, 20, 22, 23, 49
TH2, 15, 20, 22
TH17, 21–23, 40, 49
TNF-α receptor associated factor (TRAF), 14, 17, 18, 48
TNF receptor superfamily, 12, 43
TNF-related apoptosis-inducing ligand (TRAIL), 12
Torus (palatinus, mandibularis), 28
Transforming growth factor-β2 (TGF-β2), 8
Translating research into practice (TRIP), 98, 99
Translational CRCSR's (T-CRCSR's), 76, 99
Translational evidence-based osteoimmunology, 2, 99, 101, 102
Translational TRIP (T-TRIP), 99
Transtheoretical model, 90
T-scores, 50
Tubular bone, 57
Type I collagen, 3, 10, 41, 56

U
Utility model for decision-making, 74

V
Vidian nerve, 28
Visco-supplementation, 79, 80
Vit D: vitamin D_3 (cholecalciferol, derivative of cholesterol), 4, 5, 13, 16, 18, 21, 41, 44, 46, 59, 81, 84, 85
Volkmann's canals, 6
Vomer, 25, 28–30

W
Web of science, 85
WHOLIS, 88
World Health Organization (WHO), 41, 46, 50, 88, 94

Z
Zygomatic, 25, 28, 30, 61, 62